Creation

Creation

The Origin of Life

ADAM RUTHERFORD

VIKING
an imprint of
PENGUIN BOOKS

VIKING

Published by the Penguin Group

Penguin Books Ltd, 80 Strand, London WC2R ORL, England

Penguin Group (USA) Inc., 375 Hudson Street, New York, New York 10014, USA

Penguin Group (Canada), 90 Eglinton Avenue East, Suite 700, Toronto, Ontario, Canada M4P 2Y3
(a division of Pearson Penguin Canada Inc.)

Penguin Ireland, 25 St Stephen's Green, Dublin 2, Ireland (a division of Penguin Books Ltd)

Penguin Group (Australia), 707 Collins Street, Melbourne, Victoria 3008, Australia
(a division of Pearson Australia Group Pty Ltd)

Penguin Books India Pvt Ltd, 11 Community Centre, Panchsheel Park, New Delhi – 110 017, India

Penguin Group (NZ), 67 Apollo Drive, Rosedale, Auckland 0632, New Zealand
(a division of Pearson New Zealand Ltd)

Penguin Books (South Africa) (Pty) Ltd, Block D, Rosebank Office Park,
181 Jan Smuts Avenue, Parktown North, Gauteng 2193, South Africa

Penguin Books Ltd, Registered Offices: 80 Strand, London WC2R ORL, England

www.penguin.com

First published 2013
001

Set in 12/14.75 pt Bembo Book MT Std
Typeset by Jouve (UK), Milton Keynes
Printed in Great Britain by Clays Ltd, St Ives plc

A CIP catalogue record for this book is available from the British Library

HARDBACK ISBN: 978–0–670–92044–0
TRADE PAPERBACK ISBN: 978–0–670–92046–4

www.greenpenguin.co.uk

MIX
Paper from
responsible sources
FSC
www.fsc.org   FSC™ C018179

Penguin Books is committed to a sustainable
future for our business, our readers and our planet.
This book is made from Forest Stewardship
Council™ certified paper.

ALWAYS LEARNING   **PEARSON**

For David Rutherford, from whose cells I came

Contents

Foreword

The best books about evolution have probably already been written. That is testament to the grand, brilliant and demonstrably correct idea that underlies it. In November 1859, Charles Darwin published *On the Origin of Species*, in which he outlined his thoughtful arguments about evolution via the mechanism of natural selection. Although the progress of science dictates that theories and models are constantly evolving, in the 150 years since that publication, every aspect of biological research has effectively served to reinforce the central idea in the *Origin of Species*, that species change over time with the gain or loss of characteristics according to their usefulness – often referred to with Darwin's own aphorism: 'descent with modification'. A century and a half of research has shown beyond reasonable doubt quite how robust an idea natural selection is.

This book is about what came before the origin of species, and what comes next, as we engineer life forms outside of the constraints of natural selection. In Hollywood terms, it concerns the prequel and the sequel to life. Both halves are also concerned with modified descent. In this part we will trace the search for the origin of life – the ascent of lifeless chemistry into biology, in the tumult of the elemental rocks, seas and the bubbling swirl of the infant Earth. The other part explores the modification of life by human agency – the designing, engineering and building of new life forms for purpose. To explore these endeavours, we need to understand four billion years of evolution, and the two or three centuries of biology that have led to this critical point in human history. These two fields are intimately interdependent, a tangled thicket of ideas whose connections are testament to the scientific method and our insatiable curiosity to find things out. As we have come to an ever-greater understanding of the processes at the beginning of evolution, we've learned how to profoundly manipulate biology in the present, and vice versa: as we take cells apart and reassemble them synthetically we learn more about the conditions in which the first cells arose.

For that reason, this is a book in two halves, each of which can be read independently, though both stories are inextricably connected. There are two front covers, and this foreword, modified with minor suitable adaptations, appears at the beginning of both. To begin with the future of life, turn the book upside down now.

The Origin of Life

Introduction

Imagine you were unlucky enough to get a paper cut opening this book. It's an annoying but trivial injury, painful but easily mended. Yet the response that this incision triggers is complex, organized and profound. It's comparable to the human reaction to a large-scale catastrophe such as a flood or an earthquake. As in those disasters, the first phase is an emergency response.

Everything that occurs in and around your cut happens as a beautiful orchestration of individual living cells. At the precise moment the sharp edge of the paper slices through the outermost surface of your skin, cells embedded throughout your flesh called nociceptors spark into action. Via long, stringy nerve fibres that sprout from their surfaces, an electrical signal zaps from your fingertip to cells in the cortex of your brain in a fraction of a second. There, you perceive pain, and at the speed of thought your brain fires a message back to groups of muscle cells in your arm, telling them to twitch in a coordinated fashion. The muscle contracts. Your arm recoils. All of this happens within a heartbeat.

The cut will have riven cells from each other in the walls of a blood vessel, a key event in kick-starting the healing process. Opening a capillary causes blood to flood the wound. The scarlet of blood is haemoglobin, the protein molecule that ferries oxygen around your body, and it's packed into the concave discs of red blood cells shaped like a round mint half-sucked on both sides. Red cells account for just under half of the five litres of blood that the average person carries. Most of the rest is made up of plasma, which is mostly water. But in that plasma, comprising less than one per cent of blood, are cells that are utterly critical in repairing your wound. These are white blood cells, and their job is to find, fight and thwart any opportunistic invaders such as bacteria that immediately start trying to sneak into the body so that they might flourish, but in doing so cause you infection.

Meanwhile, the tip of the nerve cell that triggered that pain sends out a signal in the blood that attracts platelet cells. These are the body's 'rapid response unit', clumping together to form a clot to prevent further blood loss. They also act as emergency beacons, sending out signals to summon dozens of other 'workers' – cells and proteins that protect the wound, and initiate the process of rebuilding. Muscular cells in the artery walls spasm in synchronized contraction. Your finger throbs. This twitching restricts blood flow and loss, and helps keep immune cells where the action is. The formation of a clot prevents blood loss and haemorrhage, and marks the first phase of wound healing. Now that the barrier between the inside of your body and the rest of the world is re-established, the clean-up and restoration can proceed.

After an hour, the majority of the cells attending the paper cut are called neutrophils. These carry detectors on their membranes that pick up the chemical emergency signals pulsing out from ground zero, and move in the direction of the strongest of them. On arrival, neutrophils act as specialist cleaners, enveloping bacteria and hoovering up debris and detritus, before killing themselves when their task is complete.

Over the next twenty-four hours, another regiment of cells files into the site and each matures into the giant Pac-Man of the immune system, the 'macrophage' (literally 'big eater' in Greek). These chomp up the neutrophil carcasses and any other potentially damaging remains they find.

Crucially, the cut itself isn't simply stuck back together; otherwise we would lose the sensitivity that was there before the injury. Nor is it simply a case of plugging the gap with new skin cells, otherwise we would be lumpy and malformed. Our bodies strive to make repairs as invisible as possible, and to restore the body to its pre-injury state. It will need to be patched up with new flesh, which is a complex collaboration of cells. And that means the birth of tissue.

As with any reconstruction the foundations must be laid first. At the cut, building-block cells called fibroblasts flood the site over the next couple of days, reproducing themselves and migrating across the surface of the wound from all directions, extending ruffle-like feelers called pseudopodia – 'fake feet' – as they shuffle along. The march stops when they meet in the middle, forming a layer of foundation for the full reconstruction, and they begin to turn themselves into parts of new blood vessels and new skin tissue. Locked down in position, they produce and ooze collagen to lay down a sort of matrix or scaffold for the rest of the reconstruction project.

Skin, of course, is made of cells, but not just one type. Our skin cells grow, layer by layer, from within, with dead cells sloughing off on the outside in a process of continual renewal. Embedded within this matrix are also a host of other cells, including hair follicles, sweat and oil glands and the blood vessels that feed the flesh with oxygen and nutrients. All of these types of cell have to be rebuilt into the repair tissue.

A month after you might have opened this book, the cut is effectively healed. But long after you have forgotten all about it, the cells of your body will continue work on the wound for months, maybe a year, remodelling the site to restore it as well as it can and minimizing any scar. The redness fades as the temporary blood vessels that extended to fuel the repair process retreat, and the temporary collagen matrix that acted as scaffold for the rebuild is replaced with a more permanent version. A new piece of living tissue, a new piece of you, has been formed in an act of minor but necessary repair.

The entirety of this reconstruction project has been brought about by thousands of cells working together and producing thousands of new highly specialized cells that make up tissue: epidermis, glands, veins and arteries. The fact that we, and all life, can create new living tissue out of cells is the grand idea that unites not only all living things, but every living thing that has ever existed.

There are more living cells on Earth than there are stars in the known universe. Take bacteria: my crude estimate is that there are something like five million trillion trillion (5,000,000,000,000,000, 000,000,000,000,000) bacterial cells on Earth. You are probably about fifty trillion human cells, give or take a few tens of trillion depending on your size. On top of all that we are all walking Petri-dishes. Healthy humans are born sterile, a clean slate of only your own cells. But in a typical adult the total human cell count is out-numbered ten-to-one by the non-human cells that we carry, in our guts and on our skin and every surface inside and out, more often than not in the form of bacteria. But there's also bacteria's cousins, archaea, as well as larger and more complex hitchhikers such as yeast cells, protists and the occasional parasitic worm. Right now, you are probably carrying more than a thousand alien species about your person, most of whom you don't even notice, many of whom you couldn't live without. Mostly, these passen-gers are a fraction of the size of human cells, so by mass we're still mostly human. By number, though, at a cellular level, we are mostly other things.

These numbers are so large as to be bewildering. But they do reflect that our planet is flooded with life – and that life is only carried by cells. Though the reconstruction project of your heal-ing paper cut is awesomely complex, at the same time, it is utterly mundane: a cut, a wince, and within a few days, new skin to patch it up. Similar processes occur billions of times every minute all over our planet. Not just paper cuts and wound healing, but the fundamental processes of all life take place through cells.

As we shall see, every single one of these cells, including the new skin cells on your finger, was born when an existing cell div-ided in two. And because of that simple fact the cells in your cut have an ancestry dating back four billion years to the one excep-tion to this rule – the very first cell. When new cells are born to patch up that cut, they are the latest in a chain that leads back, per-fectly unbroken, to the very beginning of life on Earth.

So how do we know this? What is a cell? How are they able to perform this or indeed any process? How a cut heals may seem trivial, a biological act, albeit one of sophisticated orchestration, that is performed without thought. But in asking how such a refined process came to be, we will arrive ultimately at the same questions and answers that address the nature, and the origin, of life. This is a book about those questions and their possible answers: the origins of our lives, the origin of all life, and the origins of new life. And at the heart of them all is our modern understanding of the cell.

Where life comes from is one of a small handful of the most fundamental questions we can ask. It is a question that has pre-occupied humanity throughout its existence. Every culture and every religion has a creation myth, from the ancient Egyptians, who have a god sneezing, spitting and masturbating to create the world, to the comparatively restrained story of Christian Genesis, where life is created *ex nihilo* – out of nothing – and humans out of mud.

The true story began to reveal itself in a period of less than a hundred years between the middle of the nineteenth and twenti-eth centuries, with the emergence of the three great ideas in biology. As we will find out, cell theory, Darwin's theory of evo-lution by natural selection and later the discovery of the structure of DNA combine neatly to describe how life works. But they also bring us to the brink of cracking the big question itself: how life began.

Key elements of that narrative are elusive, and contemporary scientists are hard at work on filling in the gaps. With more fervour than at any time before, a model of genesis is emerging, using the weight of modern biology to reconstruct the deep past. It is only very recently, with our solid understanding of the genes, proteins and mechanics of these living chemical processes, that we can ser-iously quiz how they came to be in the first place. Modern biology has revealed complexity rather than simplicity, and shown that intricate networks of chemical reactions drive reproduction, inheritance, sensation, movement, thought and all of the things that life does. None of this happens for free: energy is required to fuel these actions. The bottom line is that, without energy, you are dead. In order to work out how life began, we will have to unpick all of these networks. It is here, in the microscopic and indeed atomic world of the cell, that we will find the clues to understand these processes – the ones that keep you, your cells and every cell alive, as they have for billions of years.

As in contemporary astronomy and physics, the great theories of biology are now being tested with groundbreaking experimentation. Our exploration of the universe led to the Big Bang theory of the origin of everything, and at the Large Hadron Collider on the Swiss–French border the biggest experiment ever undertaken has been running to re-create the universe in its most embryonic form, billionths of a second after conception. In performing this act, physicists have unveiled the Higgs Boson, a most elemental particle that features at the beginning of time and continuously since then. So, too, the best way scientists can understand the pathway to life on Earth is also to try to re-create it. In the next few years, for only the second time in four billion years, a living thing, probably something akin to a cell, will be born in the laboratory without coming from an existing cell. This is a golden age in biology, and feverish work is occurring all around the world to resolve some fundamental questions. Our journey to this revolutionary point in the Earth's history is the story of biology itself. And that story, too, begins with cells.

1. Begotten, Not Created

Life is made of cells. For that reason, the vast array of these microscopic gloopy bags is beyond description. In one single species, us for example, there isn't an absolute number, but of the fifty trillion or so that an adult might have, there are hundreds of types, from astrocytes in the brain to the zymogenic cells of the stomach. Along with that variety comes an assortment of dimensions. The longest cells are neurons in the spine, which stretch all the way to your big toe. If size is important, then you should turn to sex. Among the largest cells in humans are the eggs, just about big enough to see with the naked eye. And the smallest cells are their counterparts, sperm. Though what men lack in size they make up for in numbers: the average adult male can produce ten billion sperm per month, whereas women carry a finite number of eggs, one released every month from alternating ovaries between puberty and the menopause. Women are born with all of their eggs already in place, meaning that your first cell began its life inside your grandmother. Other than eggs, almost all cells are invisible to the naked eye, and even with a microscope most look unremarkable: tiny colourless blobs bound by a fractionally less colourless membrane, generally sitting in a fairly nondescript underlit swill. In labs, we freeze tissue and slice it into slivers less than one hundredth of a millimetre thick on glass slides, and the cells appear packed together in dense abstract patterns. Or we grow them in broth, where they can be seen free-floating like blurry stars in an off-white sky. We stain cells in hues of pink and purple, and in recent years in fluorescent greens and reds, to help visualize their inner workings. But in a live body, most are jellyfish opaque.

Each type of cell is a highly specialized member of a community, working in unison with others to build a fully functional organism. Every process of our lives is a result of those cells performing their jobs. As you read this sentence, the muscle cells around your eyeball contract and relax to control movement of your eyes from left to right. Now, if you glance above this page and look at something in the distance, a ring of muscle cells pull focus by stretching the clear cells in the lens. You move your eyes effortlessly, but this action requires intricate unconscious coordination. Photons of light pass through your lens and hit the cone and rod photoreceptor cells at the back of your eye, in your retina. There they are harvested and converted into electrical impulses that zoom through neurons, via the optic nerve up to the brain, for processing, perception and, with luck, understanding. Each movement, every heartbeat, thought and emotion you've ever had, every feeling of love or hatred, boredom, excitement, pain, frustration or joy, every time you've been drunk and then hungover, every bruise, sneeze, itch or snotty nose, every single thing you've ever heard, seen, smelt or tasted is your cells communicating with each other and the rest of the universe.

Douglas Adams once suggested that Earth was not the most appropriate name for our planet, as most of the surface is not solid dirt and rock, but water. But if you really wanted to name our home world after a feature that truly differentiates it from the other 800 or so planets that we have found, it would be Cells. Earth, uniquely as far as we know, is bursting with life, and every living thing on our planet is made of cells. Bearing in mind that nine out of ten things that have ever lived on Earth are already extinct, the number of cells that have ever existed is utterly incomprehensible.

This is a very modern understanding. Biology is a young science, at most 350 years old by any sensible count, and only 150 in terms of a fuller, mature view with comprehensive and universal rules. Physics has an older pedigree. By the mid seventeenth century, scientists had mapped areas of the cosmos with future-proof accuracy. Isaac Newton was drawing up a set of rules that explained why things move the way they do, and why we can stand on Earth and not float away. But what are now known as the 'life sciences' were a long way behind. The reason for this is that the starting point for most scientific advances is to look at things, and work out why they are the way they are. Unlike the stars and planets, no one had even seen – or at least identified – a cell before 1673.

At this time, science itself was forming. Gentlemen scientists such as Newton and Robert Hooke had founded the world's first scientific body, the Royal Society. But the man who first peered into the minuscule world of the cell at the birth of cell biology was not one of the esteemed bewigged gentlemen of science. The unlikely beginning of the story of biology must be credited to a Dutch linen merchant called Antonie van Leuwenhoek.

The business of making and selling cloth was inextricably linked to the development of better optical lenses, as merchants checked the density of fibres and therefore the quality of their fabric using magnifying glasses like a watchmaker's loupe. Van Leuwenhoek was a skilled and meticulous lens-grinder, working in Delft, the capital of Dutch drapery. He specialized in a technique that involved pulling apart a hot glass rod and squashing the ends back to form a ball, but was secretive about this process, as it had made him the greatest microscopist of his day. Van Leuwenhoek's lenses were tiny fat drops, not much bigger than a peppercorn, and he attached them to handheld contraptions nothing like the microscopes that are familiar to us today. His were rectangular copper plates, about an inch by two inches, with a hole at one end to hold the rotund glass bead lens. On one side there was a silver spike to hold the specimen in front of the lens, held by a screw that could be turned to pull focus. It was their fatness that gave van Leuwenhoek's lenses their superior magnification.

At least, that was his technological advantage. His other key attribute was insatiable curiosity. Simply, van Leuwenhoek liked looking at small things through his lenses. While I hope the paper cut described in the introduction is entirely imagined, van Leuwenhoek deliberately invoked exactly the same repair process out of unbridled curiosity. In a letter published in the Royal Society's official journal *Philosophical Transactions* in April 1673, van

Leuwenhoek wrote: 'I have divers times endeavoured to see and to know, what parts Blood consists of; and at length I have observ'd taking some blood out of my own hand, that it consists of small round *globuls*.' We think that he was looking at red blood cells, and this appears to be the very first recorded sighting of individual cells.*

As his microscopy skills improved he began to look at all manner of bodily samples and fluids. He went on to scrape the matter from between his teeth, and observed the bacteria that cause plaque. At the tail end of the seventeenth century, van Leuwenhoek was becoming something of a celebrity for his exploration of a microscopic kingdom hidden in plain view. King William III of England and other dignitaries visited him to see what he had seen. One investigation, however, was kept private: his own semen, although he attested in his notes that the sample was acquired 'not by sinfully defiling myself, but as a natural by-product of conjugal coitus'. In this act, which it is perhaps best not to dwell on, he saw sperm for what they are: single cells. He also discovered cells in a bead of water from a local lake and saw what we now loosely call 'protists': single-celled creatures that include self-powered swimmers, and algae.

Van Leuwenhoek was the first person to definitively see individual red blood cells, sperm, bacteria and free-living single-celled organisms. This last group he gave a cute name, 'animalcules', and in the 1670s sent drawings of his discovery to the Royal Society in London. The fellows expressed scepticism, not least because, when they asked Robert Hooke, their resident microscope expert, to see if he could see the same creatures in water from the Thames, he initially saw nothing.

Hooke's expertise in looking at tiny things was unparalleled, having published a stunning and popular volume a decade earlier: *Micrographia: or Some Physiological Descriptions of Minute Bodies made*

---

*I say 'think' because he also describes seeing *globuls* in milk, and these are likely to have been suspended droplets of fat.

by Magnifying Glasses. Rarely has a book been so accurately sub-titled. It contains, as you might expect, annotated drawings of very small things. His microscope was a simple six-inch tube with two lenses, and a cricket-ball-sized crystal sphere to magnify the illu-minating flame. Many of the images generated by this kit are now very familiar, including a giant fold-out of a flea, and a terrifying close-up of the eyes of a hover fly, which looks staggeringly simi-lar to contemporary photos taken with an almost unrecognizable descendent of Hooke's tools, the electron microscope. Samuel Pepys picked up a copy of *Micrographia* and noted in his diary that it was 'the most ingenious book that ever I read in my life'.

Quite right. But there's a significant irony contained within this magnificent tome. One of Hooke's detailed illustrations is of a longitudinal cross-section of some bark from a cork tree. There in the meticulous picture are conjoined units making up the overall structure. In the text, Hooke uses the term 'cell' to describe these units. In reality they are in fact the dead walls that once housed cork tree cells. He chose the word as it derived from the Latin *cella*, meaning cubicle, but noted that they were air-filled, which helped explain cork's buoyancy. Hooke had seen the remnants of cells, named them cells, but did not for a moment imagine that his observations were of universal living units, nor the permanent legacy of the name.

That is how cells were discovered. But where did they come from and how were they formed? With curiosity and technology, van Leuwenhoek had pulled back the curtain to reveal an undis-covered kingdom, but progress was fundamentally hindered by a resolutely unenlightened ideology.

The Origin of Cells

The origin of the cells van Leuwenhoek was glimpsing remained elusive. Obviously, people knew that sexual intercourse between a male and a female often resulted in a new life, but, perhaps because most sex goes unobserved, an entirely fanciful view of the origin of cells remained vexatiously persistent.

'Spontaneous generation' was the most popular origin-of-life story for thousands of years. The first substantive explanation of spontaneous generation comes from Aristotle, the unsung father of biology. In his book *Animalia* (The History of Animals), written in the mid-third century BCE, he describes the genesis of certain species:

> So with animals, some spring from parent animals according to their kind, whilst others grow spontaneously and not from kindred stock; and of these instances of spontaneous generation some come from putrefying earth or vegetable matter, as is the case with a number of insects, while others are spontaneously generated in the inside of animals out of the secretions of their several organs.

Animalia is a fascinating book, probably the first great biology textbook. It's crammed with observations and conclusions about a huge range of species, some of which are dodgy.*

In spontaneous generation Aristotle described an idea that persisted until the nineteenth century, during which time dozens of examples materialized. The Roman writer Vitruvius casually referred to spontaneous generation when advising architects in the first century BCE:

> libraries should be towards the east, for their purposes require the morning light: in libraries the books are in this aspect preserved from decay; those that are towards the south and west are injured by the worm and by the damp, which the moist winds generate and nourish, and spreading the damp, make the books mouldy.

---

* For example, Aristotle declares that some fish are neither male nor female, and these are prone to spontaneous generation unlike their sexed brethren. We now know that all fish are sexual beasts, to the extent that many such as clownfish and wrasses can switch sex when their environments demand it.

And in the sixteenth century Ziegler of Strasbourg claimed that the origin of lemmings was that they descended in their hordes from storm clouds.*

Indeed, following van Leuwenhoek's discoveries, all manner of speculation on the origin of cells and life was put forward, all involving spontaneous generation. In seventeenth-century Brussels Jean Baptiste van Helmont – marked in history as a respected scientist, the father of the chemistry of gases – described an experiment in which he placed a sweaty shirt in a vessel with some wheat and let it ferment in his dank castle basement for twenty-one days. Lo and behold, this concoction gave rise to mice.†

It's all too easy to mock the ignorance of the past, and we should be wary of it. Though spontaneous generation was an obstinate idea, it is not a solid scientific concept. All the misplaced examples, most notably when describing the genesis of larger beasts, were born of incomplete or poor observations. There are few more elementary questions than that of the origin of life, either in terms of the origin of new life from existing life or the more fundamental absolute origin of life, *ex nihilo*.

It was almost 200 years after van Leuwenhoek first saw cells before this idea would be put down. The final dismissal of spontaneous generation bears all the hallmarks of good science: rigorous observation and a testable, predictive hypothesis. But it also comes with the stamp of pure drama: an international cast and a generous helping of money, fame and betrayal.

---

*The notion that lemmings are suicidal is equally fantastical, a myth derived from images of mass migration, possibly staged, in the 1958 Disney film *White Wilderness*.

† As every biologist who has worked with them knows, gestation for the common house mouse is indeed around three weeks, but, more importantly here, the simple fact is that mice are sneaky, small and hungry: where there is grain, you'll find mice.

The Birth of Cell Theory

The quality of microscopes improved steadily over the eighteenth and nineteenth centuries, and the popularity of the study of small things grew accordingly. The biggest advances were coming not from exploring the microscopic animal kingdom, but from looking at plants and algae. That different parts of plants were composed of cells was apparent in the first few decades of the nineteenth century, though the ubiquity of them in all living things was not. Much of this work was carried out in Germany, and the textbooks carried names describing a range of observed structures: Körnchen, Kügelchen and Klümpchen (granules, vesicles and blobs). But while the description of tissues was progressing, the root of its genesis was not. It was not until 1832 that the birth of cells was first described. A Belgian baron called Barthélemy Dumortier watched cells in an algae grow longer and longer until a partition wall appeared and one cell became two. Others soon replicated the work and observed it in different algae and plants.

Even without a consistent model for reproduction, the innards of cells were being explored. The quality of microscopes was increasing, and in 1831 Robert Brown was studying orchid cells.* Within them he observed 'a single circular areola, generally somewhat more opaque than the membrane of the cell'. He called it the nucleus, the name it still bears today, and we now know it as the central office for the genetic code in all complex organisms.†

As with Dumortier's observation of cell division, Brown's

---

*Robert Brown is most famous for Brownian Motion, which describes the haphazard microscopic pathway that particles make in gas or solution when bombarded by atoms of molecules.

†While Brown certainly named it, the prize for the first observation of the nucleus goes, yet again, to our Dutch friend Antonie van Leuwenhoek. In a letter to Robert Hooke dated 1682, he describes smaller bodies within the red blood cells of a fish. Admittedly, it's an incomplete description, and doesn't imply any of the subsequent importance of the nucleus. But an unnamed modern

assumption was that the nucleus was not universal. Many thought division was an exceptional means of cell birth, and no such observation of fission had been seen in animal tissue. Plant cells are frequently considerably larger than their animal counterparts, and so the study of flesh trailed the examination of leaves. The nucleus had been observed in some animal tissues, particularly brain cells, but again, was not assumed to be in all cells. Compounding the issue, the single most common type of cell in humans, red blood cells, do not contain a nucleus: it is shed during their development.

The names most closely associated with cell theory in almost every textbook are Schwann and Schleiden. The story that Theodor Schwann told was that there was a 'eureka' moment at the birth of cell theory. Schwann and Matthias Schleiden met by chance at a dinner in 1837. Schwann was an anatomist, short and nerdy, and would sometimes not leave his digs for days at a time while he examined bodily tissue. Schleiden was a troubled and sometimes suicidal botanist who had been influenced by Robert Brown's identification of the nucleus. Botany and animal biology were separate fields, far from the unification that evolution and genetics would ultimately supply. Over dinner they discussed their work, on animal and plant tissue, respectively. The conversation must have become even more thrilling for the other guests as their discussion progressed to the nucleus, that smaller body in the middle of cells. Schwann and Schleiden both realized that the nucleus was the same in plant and animal cells. They rushed without delay to Schwann's laboratory to compare notes. From this point on, the idea that all living tissue was made of cells took hold.

Dynamic though this legend is, Schwann and Schleiden were in fact only partial contributors in the development of a model of life based on cells, and significantly wrong in one major part. Much of the work demonstrating the nucleus in plants and animals had

---

editor at the Royal Society, where these letters are kept, has casually scrawled in the margin 'Discovery of the cellular nucleus'.

come before them, and the universality of cells had been suggested by others before 1837. It was probably Schwann who first used the phrase 'cell theory', but where he and Schleiden were most significantly wide of the mark was on the origin of new cells. Both described the formation of new cells starting with the spontaneous appearance of a naked nucleus in the spaces between existing cells. According to them, that nucleus acted as a seed from which the new cell would emerge, like a growing crystal. It's not as far-fetched as heavenly lemmings, but otherwise it was still vitally reminiscent of spontaneous generation.

Our understanding of the origin of new cells can be largely attributed to Robert Remak – a lost hero of biology, and a victim of politics and race. Remak was a Polish Jew who spent his adult life in Berlin. To get the university position he wanted and deserved, he would have had to betray his orthodox Jewish roots and be baptized, something he never did. Through his undeniably good science, he was eventually given a post of lecturer and subsequently assistant professor at the University of Berlin, but this post came with neither salary nor lab. Compare that to his contemporary cell biologist Rudolf Virchow. Born of a well-to-do Prussian family, Virchow was flamboyant and bombastic. Ultimately he would be described as 'the pope of medicine' and 'the sole instance of a full-fledged physician-scientist-statesman in our time'. He was six years Remak's junior, but they were appointed to the University of Berlin at the same time.

After careful observation, Remak rejected spontaneous generation in all forms, including that which Schwann and Schleiden were describing. Over a decade, he had studied all manner of animal tissues, including muscle, red blood cells, and in frog and chicken embryos, and he only saw cells splitting, one pinching in the middle like a belt on a balloon until it became two. Virchow followed Remak's work, and year by year edged closer to his way of thinking – that cells were only formed by cell division.

Then, in 1854, Virchow declared that 'there was no life but through direct succession', and a year later translated this into a

Latin motto: *omnis cellula e cellula* – all cells from cells. He was popular and prominent, and expounded this idea wherever he could, including in his international best-selling textbook *Die Cellularpathologie*. But there was no mention of Remak in any of these writings. Virchow had done little of the work, but had adopted his colleague's toil without any credit. Remak was livid and wrote to Virchow about the Latin summary:

> [it] appears as your own without any mention of my name. That you make yourself ridiculous thereby in the eyes of the knowledgeable, since you have no evident embryological expertise, neither I nor anyone else can undo. If, however, you wish to avoid a public discussion of the matter, I would ask you to immediately acknowledge my contribution.

We sometimes forget that science is done by people, with all their personalities in tow. The scientific method is designed to equilibrate for all those idiosyncrasies and ultimately does tend to. But due credit is a perpetual problem.*

Nevertheless, Remak, and Virchow for his sins, had nailed it. Life is cells, and cells are only begotten from other cells. But, like a zombie, spontaneous generation still shambled along, lurching up yet again in France in 1860. The man who finally killed it dead was Louis Pasteur. Not yet famous for the sterilization technique that bears his name, Pasteur was ambitious and young, and had twice been denied membership of the French Academy of Sciences.

An experiment by a prominent proponent of spontaneous generation had reignited belief in its existence. Felix Pouchet wanted

---

* For the record, Virchow was not a bad guy, even though this looks like the act of a backstabbing rat. Throughout his career he was tremendously politically active, fought social injustice and drove civic reform in Germany and Prussia with great success. According to one story, his liberal views annoyed the Prussian prime minister Otto von Bismarck enough that he challenged Virchow to a duel. As the challenged, Virchow had the right to choose weapons. He selected sausages, one cooked, the other loaded with roundworm. Bismarck, the sausage-fearing Iron Chancellor, withdrew.

to show that mould would spring forth from hay, even if the hay, the air and the water used were sterile. Pouchet had boiled the ingredients and then cooled them down with liquid mercury. As if by magic, mould appeared on the hay. The Academy wished to sort this out once and for all, and offered up a prize of 2,500 francs to the first person to resolve spontaneous generation.

Pasteur spotted the flaw, which was that the mercury had a layer of dust on its surface, which was bringing in the mould. So he designed the simplest experiment imaginable. His version of Pouchet's set-up was to have two flasks containing a sterile but rich broth, one that would soon go cloudy if exposed to microbial life. One flask was left open, while the other had an s-shaped curved neck protruding to one side. Pasteur figured that microbes would reach the broth in the first flask, carried on dust particles in the air, but that the swan neck would not allow these contaminants into the broth in the second flask.

Within days, the open flask was cloudy. But the swan-necked flask was perfectly clear, and indefinitely so. As a control, Pasteur would snap off the swan neck and observe the broth grow cloudy over the next few days. Pasteur claimed the cash, and was duly elected to France's scientific elite.*

The ultimate fate of this obstinate idea is best described with Pasteur's own words: 'Never again will the doctrine of Spontaneous Generation recover from the mortal blow struck by this simple experiment' and 'those who think otherwise have been deluded by their poorly conducted experiments, full of errors they neither knew how to perceive, nor how to avoid'.

These are harsh words, but true. Thousands of years of biological superstition were swept away by that essential feature of science: the experiment. And with that swan-necked flask, cell theory was completed. Like all great theories, it's a fusion of ideas,

---

* To further prove his point that contaminants were in the air, he repeated this in different locations, some in dusty rooms, others in the relative sterility of 800 metres altitude on Mont Blanc. The results were consistent: clean air, no growth.

based on observation and affirmed by experimentation. It is also one of the great nodes of biology. The work of dozens of men and hundreds of years of investigation into the stuff that life is made of is summarized in two sentences:

1) All life is made of cells.
2) Cells only arise by the division of other cells.

This theory has profound implications, as all grand theories should. It covers all life, a simple but comprehensive description of the innumerable inhabitants of the living Earth. But, as we already know, there are trillions of different types of cells. In the case of red blood cells, for example, those found in humans are different enough even from those of our nearest primate cousins that they cannot be exchanged without serious consequences. By the time we start looking at species as distant from us as chickens we find, as Remak did, that they do contain a nucleus, unlike our own red blood cells. The first grand idea showed that the diversity of life on Earth was embedded in the magnificent range of cells. The second showed how that diversity came to be in the first place.

Change Over Time

At about the same time as Schwann, Schleiden, Remak and others were observing how cells behave, across the English Channel a youngish family man was busy reflecting on his over-extended gap year. Charles Darwin was slowly and meticulously putting together an overwhelmingly compelling case that describes how creatures evolve. Evolution, the idea that species are not immutable, was already being contemplated as a concept in the nineteenth century. But the process by which they change was unknown. Darwin had spent five years travelling thousands of miles on HMS *Beagle*, collecting specimens from the far side of the world. On returning, he married his cousin Emma Wedgwood: they were both grand-children of the pottery magnate Josiah Wedgwood. They settled in Down House in Kent, and, untroubled by financial pressures, Darwin began chiselling out a splendid idea. In 1859, after years of scientific and personal struggle, Darwin published *On the Origin of Species*.* In it he sets out the second grand unifying theory of biology, one describing the process by which evolution takes place.†

Unlike the microscopists in Europe, Darwin was primarily concerned with whole animals, the macroscopic world. He observed that, when comparing individuals in any population, there is a natural range for any given physical characteristic. This variation is the means by which those individuals can have a competitive advantage over others. In an imaginary population of anteaters, one with a slightly longer tongue than his contemporaries may be able to root out more juicy termites, and may be better fed and healthier as a result. This could have the effect of making him live longer, or maybe a more attractive mate to a female anteater. Consequently, this anteater may have more baby anteaters, each potentially carrying that longer tongue. After a few generations of that success, because that tongue is a useful thing, long-tongued anteaters may come to dominate the population and become the norm. Over successive generations the species will change. In contrast to earlier theories, Darwin observed that it was not characteristics acquired during life that were passed on to offspring. Years of obsessive measurements led him to establish the

---

* Darwin was ultimately prompted to publish by another explorer biologist, Alfred Russel Wallace, who came up with effectively the same idea, and wrote to him describing it. Darwin, a true gentleman, suggested they went public with their idea together.

† As a result of Darwin being a scrupulous note-taker, and writing down pretty much everything he saw and did, his important life has been documented in meticulous detail. There is an ongoing leviathan project to digitize everything he ever wrote, scribbled, jotted and doodled, and publish it electronically at The Complete Work of Charles Darwin Online. There you can read about everything from his experimental bassoon-playing to earthworms to the wooden slide he had built for the central staircase in Down House to keep his many offspring entertained. As a result of his coming up with pretty much the best idea anyone ever had, the body of literature about evolution is deservedly both gargantuan and wondrous.

principle that for each and every trait – an anteater's tongue, hair colour, any trait you care to name – it was the variation of that characteristic across the population that conferred advantage for an individual. Those traits would become more common in a population because the advantage they bestowed would play out as greater reproductive success.

There are other significant selective forces, such as the complexities of sex, where males big themselves up to win a female, or females exercise seemingly whimsical choice over males. But natural selection is the overarching force that has shaped the living world in which we live. It's a system of error, trial and revision. Evolution is blind, and has no direction. Species are not more or less evolved, nor are they higher or lower, as they were once and sometimes still are described. Through iteration, they are merely better adapted to survive in their environments. In an experiment that really should not take place, an orang-utan wouldn't last two minutes in the boiling submarine waters of a hydrothermal vent, despite being sophisticated enough to use tools in the jungles of Borneo. Down in the hot sea of a vent though, hundreds of species, including the two-metre giant tubeworm and dozens of species of bacteria, are quite content to eke out their existence. Change is the norm, and adaptation is success.

In the 150 years since the *Origin of Species* was published, millions of scientists have poked and pulled at the theory of evolution, unpicked, tweaked and yanked at it in every conceivable way. They have observed countless species from aardvarks (or anteaters) to zebras to study their behaviour. They have created simulations of myriad populations first with mathematical models, and later in computers, and pushed and pressurized their artificial environments to see how they adapt over successive generations. They have bred and crossbred inestimable species to observe how inheritance works, to see where advantage lies in the next generation. They have let tanks of bacteria reproduce for decades and witnessed descent with modification in action. In the modern era, we have deciphered their genetic codes and seen exactly the differences in DNA that reflect one species becoming two, each finding a niche into which they are better suited. We've seen populations of bacteria adapt to the hostile action of antibiotics and, distressingly, emerge resistant. While the initial model that Darwin laid out has been modified and fleshed out, the 'one big argument', as he described it, has survived intact throughout the necessary battering that an idea of this magnitude demands. This is why it is called the Theory of Evolution by Natural Selection. The colloquial meaning of 'theory' as a hunch, or guess, or plain old stab in the dark, is woefully puny next to its scientific meaning. When scientists talk about a theory they are aiming at the top of the pile of ideas: a set of testable concepts that all point to and predict a description of reality that is so robust that it is indistinguishable from fact.

Darwin drew up his masterpiece as the study of cells was beginning to emerge from the stagnant pond of spontaneous generation. But his description of evolution is not about the beginning of new life; it is, as is implicit in the title, about the origin of new species. These arise when an organism has acquired so many mutated traits that they are no longer capable of reproducing with what were once their kin. When we are taught natural selection, typically we refer to prominent visible traits – antlers, hair colour and, once in a while, anteaters' tongues. But we now know how biology works at a cellular level, and can transpose evolution by natural selection to the microscopic world that Darwin was largely unaware of. The fictional anteater's tongue is longer because, by the chance of variation within its population, that individual has more (or possibly larger) cells in that tongue, and the genes that generate this disparity of tissue will be passed on through sperm or egg to the next generation. Similarly in your paper cut, the reason your platelets form a plug and a wedge to help stop blood loss is effectively because creatures carrying cells in their blood that didn't perform that clotting function as effectively have been deselected by nature thousands of generations (and species) ago (probably by allowing the creature concerned to heal less efficiently, or possibly leak to death). Crucially, we now know that what is being selected is not the individual, nor the cell, but the carrier of the information that bears advantage. As with all biology, the information that confers clotting is held in DNA inside cells, the molecule that will play the central role throughout the whole of this story.

Cell theory and natural selection are reflections of the same truth: life is derived. It's incrementally and ultimately spectacularly modified, but, in essence, life is the adapted continuation of what came before.*

Public acceptance of evolution by natural selection has waxed and waned over time, and was disputed by scientists in its first fifty years or so. But now, at least amongst scientists and those who generally understand it, natural selection stands robustly as the primary valid explanation for the variety of life on Earth. While science by definition expects to be corrected over time, it now seems extremely unlikely that Darwin's idea will be replaced wholesale. When you tie it together with cell theory, both ideas are compellingly fortified.

Although the concept of evolution – simply that organisms change over time – predated Darwin, in 1859 these were new ideas, and truly revolutionary. They both rebutted the prevailing view

---

* Darwin's idea has, over the years, really upset a lot of people, despite being self-evidently true, demonstrably true, experimentally true and on all counts robustly so. Opposition to his radical but thorough argument was immediate and forthright, but so was support. 'How extremely stupid not to have thought of that,' said Thomas Huxley, Darwin's most pugnacious contemporary defender. It's a flattering sentiment, but masks the fastidious detail and stack of work that Darwin put into his opus. Conversely, cell theory appears to have upset exactly no one. It was observed, refined and then has simply remained true. The laws of natural selection, the laws of genetics and the workings of DNA (both of which we will come to shortly) are taught, fittingly, as absolute cornerstones of biology. But the principles of cell theory are simply assumed, and assumed to be correct. It's an odd oversight, but, I suppose, one that we shouldn't complain about.

across the majority of human history: that creatures were each created distinctly. Without the hard-fought work of Darwin, and the closely observed work of the eighteenth- and nineteenth-century microscopist, the idea of multiple routes for life, separate origins not just for plants, animals, fungus, but for every single type of organism, might have seemed plausible. Even taking into account the innumerable variety of distinct cell types, each with highly specialized functions, separate or plural origins might seem reasonable.

Instead, thanks to Darwin and cell theory, we can link every organism on the grandest pedigree. As Darwin writes in the final paragraph of his masterpiece: 'There is a grandeur in this view of life, with its several powers, having been originally breathed into a few forms or into one.' These are some of the finest words set to paper, and are oft quoted. Some things are just worth repeating. But he posed a question in those last words: 'into a few forms or into one'. Which is it? What lies at the base of the tree of life? A single form, a cell, or many? The answer to this deep-rooted historical question lies not in the past, but in the molecular guts of every single living cell. By examining the mechanism by which cells pass on their characteristics and by which those characteristics mutate, we will arrive at an answer to the question of whether life has a single origin – and we will begin to see what it might have looked like when it emerged.

2.   Into One

Look out of the nearest window and try to count how many species you can see. As I sit here at my desk gazing into the garden, I can count forty species of plant (at least six of which I can name), a spider, a squirrel and a suspiciously well-fed woodpigeon. I know that there are millions of other beings nestling in the soil, feeding on the leaves, in the mortar between bricks. I know that there are tiny parasitic insects hitching rides on the hairs of the squirrel and in the feathers of the bird, and those bugs themselves will be serving as a rich habitat for thousands of bacteria. I know that on Earth bacteria outnumber and outweigh all other living things by count and by mass, and there are billions in every spade-full of loam. As undergraduates, we grew cultures of bacteria taken from swabbing our teenage skin and streaking them out in their multitudes on agar plates. You can look down at the end of your nose and be confident that there is more life in that view than in the rest of the known off-world universe. We have classified around two million living species, and we know that number is a massive underestimate, as new ones are discovered and named every day. Life is bewilderingly diverse. What is more confounding, though, is its conformity.

It doesn't take a great leap of faith to see that we are closely related to chimpanzees or gorillas. Only a designer setting out to deceive or fundamentally lacking in imagination or effort would create things so similar and pretend that they were not cousins. It is obvious that dogs in all their human-crafted forms are related closely to wolves, and science confirms it. It takes a little bit more analysis to show that dolphins are much more closely related to hippos than to tuna, whom they superficially resemble. Dig deeper and it becomes obvious that all three have common ancestry visible in their spines, eyes and the bone structure (amongst many other things) of their flippers, feet and fins, respectively. But it's admittedly harder to see the commonality between the leaf of a sycamore and the fin of a cichlid. Or a western long-beaked echidna and the recently discovered Malaysian mushroom *Spongiforma squarepantsii*. Or *Turdus philomelos* and *Candida albicans*: one is a pretty songbird, the other a creamy fungus, but both are called 'thrush'.

This zoo is made of cells of many different sizes and shapes. And yet all cells are basically the same, in the same way that all petrol-powered cars are basically the same. Cars have a chassis, an engine, some wheels (mostly four), a steering wheel and so on. The details of the engine, the design, the tyres and so on make up the difference between a Porsche 911 and a Trabant. But fundamentally they are both cars, derived via many iterations from a common ancestor, one bearing an early version of the internal combustion engine built from metal and powered by fossil fuel. And so it is with cells. Under the bonnet, there are mechanisms and parts that vary the performance of a cell depending on its job as part of an organism or standing alone. In terms of their overall structure they are basically the same, but highly specialized to build the complexities and adaptations of all life.

The starting point for those basics of biology was the result of the cell scientists striding forward in the mid 1800s. This was clearly a feverish time as Darwin was also deducing evolution, and, a few thousand miles east, an Austrian holy man was planting a garden that would invigorate biology for evermore. Joseph Gregor Mendel is invariantly described as a monk, which, whilst absolutely true, shrouds the fact that his legacy is unequivocally as a scientific genius and world-changing experimentalist.* Around the time Darwin was writing his masterpiece, Mendel had been studying pea plants, and was breeding them in their tens of thousands. As any scientist will tell you, large numbers makes for good

* Conversely, though Mendel went on to became Abbot, history does not record how good he was at monking.

statistics. What Mendel found in impressively large numbers was that, when crossing variants of pea plants with one another, the outcomes in the offspring were entirely predictable. He showed, moreover, that the traits were inherited in a discrete way – that is, independently of the rest of the plant's characteristics. Rather than a blending of flower colour, the progeny of a purple-flower pea plant and a white-flower pea plant were not pinkish, but a predictable number of white ones and purple ones. By repeating these cross breeds until the patterns were clear, his pea experiments also resulted in his determining that characteristics were inherited equally, one from each parent, but that some of those characteristics were more equal than others. He bred tall plants with short ones, and their offspring were always tall, rather than an average of the two heights. When he crossed those offspring together, three-quarters of their offspring were tall and one was short. In those proportions, he had uncovered not only that characteristics were passed down individually, but also that some characteristics were dominant over others.

The story of Mendel and his peas is high-school biology. What he had discovered (though the name came much later) was the existence of genes – discrete units of inheritance.*

Having been largely ignored, at the beginning of the twentieth century Mendel's papers were rediscovered. What followed was

---

*Mendel performed his key experiments in a seven-year period between 1856 and 1863, and in 1866 published them in a minor parochial journal, *Proceedings of the Natural History Society in Brünn*. Mendelian scholars have determined that 115 copies of the paper were distributed, and many writers have described how one of them made its way to Darwin himself, and how it was discovered in his library after his death in 1882. But, oh, the heartbreak of it all! The pages of this manuscript were uncut. Darwin never even opened it. Imagine the fusion of these great ideas, the two great biological geniuses finally uniting their work into one all-encompassing law, the mode and the mechanism of inheritance . . . Alas, this story is fully untrue, a 'what if . . . ?' myth. There was no copy of Mendel's pea paper in Darwin's library, cut or uncut. Indeed, according to the custodians of his library, Darwin had no published work authored by Mendel in his vast collection at all. Where this charming fiction has come from is unknown.

observation beyond that which is visible to the naked eye. New technologies of the twentieth century meant that the scale of biology was reducing from the organism to the cell, to the molecular and atomic level, and with this zooming-in came the birth of modern genetics.

'It has not escaped our notice . . .'

Between Mendel's death in 1884 and the 1950s, there were major successive advances in the study of genes. Mendel had established that inheritance occurred in discrete units. Italian marine biologists looked at the cells of sea urchins and observed chromosomes – neat structures inside the nucleus of all cells, which became visible, resembling tiny sausages, when the cells divided. They came in specific numbers depending on the host, and they discovered that altering the number of chromosomes resulted in abominations in their offspring, or prevented reproduction altogether. In the 1920s, Thomas Hunt Morgan inbred fruit flies to show that Mendel's units of inheritance were positioned very precisely on these chromosomes. German researchers, meanwhile, had shown that chromosomes were made of a molecule called DNA, whose chemical composition was clearly different from the proteins that made up much of the cells' ingredients, as it contained phosphate.

In the 1940s, Oswald Avery, Colin MacLeod and Maclyn McCarty demonstrated that it was this DNA that conferred characteristics and passed them on by performing a neat resurrection trick in New York, perversely a fatal one. They were following the work of Fred Griffith, a British medical officer who more than a decade earlier had noticed that, during the course of pneumonia, virulent and benign bacteria were both present, but that the latter could acquire the malign characteristics of the former, even if these malign bacteria were themselves dead. He demonstrated this by boiling the lethal bacteria, thereby killing them, and then adding the resultant broth to the benign bugs, which then acquired the ability to kill. Avery and his team repeated Griffith's experiment, but to figure out how it worked, they systematically eliminated each of the components of the bacteria that were candidates for passing on this characteristic. It was only when they destroyed the deadly bacteria's DNA that the transformation failed to occur. It appeared that DNA, not proteins nor anything else in the cell's milieu, was the essential genetic material, the stuff that bestowed characteristics and passed them on.

DNA was clearly a – possibly *the* – key component of inheritance. But it was unclear how this could work. The answer lay in its construction. At King's College in 1952, a group of researchers including Maurice Wilkins, Rosalind Franklin and Raymond Gosling were investigating DNA using their expertise in producing photographic representations of three-dimensional molecular structures. X-ray diffraction, a standard technique for establishing the shape of complex molecules, was brought to London by Wilkins from the Manhattan Project, which created the first atomic bombs. The principle is similar to the kind of silhouette portraits that were fashionable in the eighteenth and nineteenth centuries: shine a beam at your subject, and capture the light and dark that is projected past it. The human subject in portraiture is a solid to the visible light that this technique uses, but in the molecular version the X-rays penetrate the molecule under scrutiny and create signature shadows behind it, regular but cryptic swirls cast onto the photographic plate. Mathematical deduction is required to figure out the arrangement of atoms that could produce such a pattern, but the effect is the same: a unique portrait of a molecule otherwise too small to see. Franklin was particularly skilled at this technique, and of the many photographs that she, along with Gosling, developed whilst performing this arduous method, Photo 51 would be the key to one of the great achievements in human history.

The Cambridge scientists Francis Crick and James Watson acquired the photograph. Science always builds on the work of others, but it was with their insight and genius that they deduced from Franklin and Gosling's photo that DNA took the form of a twisted ladder, the iconic double helix. In a brief paper in the scientific journal *Nature*, 25 April 1953, they showed that the rungs of this twisted ladder contained paired chemical letters – A for adenine, T for thymine, C for cytosine and G for guanine. Each letter is bound to one upright of the ladder, and pairs up with a corresponding letter on the other upright to form a rung. It is this pairing that makes the helix doubled, and the pairing is very precise: A always pairs with T and C always pairs with G. Crick and Watson concluded the paper with one of science's great understatements: 'It has not escaped our notice that the specific pairing we have postulated immediately suggests a copying mechanism for the genetic material.'

This is the first marvellous thing about DNA. If you split the double helix into its two component strands, you immediately have the information to replace the missing strand: where there is an A, the other strand should have a T, and where there is a C, the other strand needs a G. Therefore, DNA possesses an ability, inherent in its structure, to provide the instructions for its own replication. With Crick, Watson and Franklin's result, we were given a molecule that could be copied and passed from generation to generation.*

The letter molecules, known as nucleobases or simply bases, bind to one another, holding the two struts of the ladder together. There are millions of such pairings on each strand of DNA, something like two metres of it in every human cell that has a nucleus, though it is tightly wound up, and then wound up on itself again around small lumps of proteins like beads on a string. And this winds up on itself again to form a chromosome, like a thick tug-of-war rope.

The number and length of chromosomes varies hugely between all species, and that variation doesn't appear to relate to either the size or complexity of the host. We have twenty-three pairs, and

* Although we tend to think of DNA as the elegant double helix, in action inside cells, it is very dynamic, constantly shuffling, twisting, splitting, folding and remoulding as it performs its myriad duties. When extracted in labs for experiment, DNA resembles pale, stringy snot.

bacteria tend to have just one, in a neat loop. But some carp have over a hundred, and this doesn't come anywhere near the number in some plant chromosomes, which can be in the thousands. The full collection of an organism's DNA, packaged into chromosomes, is called a genome. In humans, those twenty-three chromosomes, our genome, contain around three billion of those little letters, A, T, C and G, which is sufficient to fill a 200,000-page phone book, if that type of comparison is appealing. But every time a cell splits into two, from the first time your fertilized egg divides to the new-born skin cells in a paper cut, all of the DNA in those cells copies itself. The new cell contains all the same DNA of its parent.*

Being able to replicate from generation to generation is a neat trick indeed. But alone, that would just fill the world with a pretty molecule. DNA's secret power is that this string of chemical letters, the bases, are a code that harbours information. That information is an instruction manual for all of the processes of life, including the very instructions required for the replication process itself. Understanding how DNA's code works will unlock not just how mutation and variation take place, drawing us closer to an answer to the question posed in Darwin's 'into few forms or into one', but will give us essential clues as to how life was formed in the first place.

How DNA Works

All life is made by, or of, proteins. They form the structures and catalysts of biology, and the manufactories of bone, hair and all the bits of a body that aren't actually made of protein themselves.

---

* Biology is full of caveats and exceptions, and the exception here is a rather important one. During the process of sperm or egg production, a slightly different form of cell division occurs, called meiosis. This results in cells with half the total genetic material. There is a return to the full complement when sperm meets egg at conception.

And of course this isn't limited to us, or even mammals. Every leaf, strip of bark, reptilian scale, horn, fungus, feather and flower is made of or by proteins. These workhorses of life are themselves made up of strings of smaller units called amino acids – a generic name for a potentially infinite number of molecular parts that qualify for this chemical moniker.

We now know that each gene – those discrete units of inherit-ance – in a genome is a piece of code made up of the specific pairings of DNA bases that encrypt the construction of a protein. However, only a few parts in many species' genomes are genes. The rest – in fact in humans the overwhelming majority of DNA – comprises notes, instructions, scaffolding and even insertions from interloping viruses. Some are the remnants of genes from our ancestors whose function has been lost in us, but whose ghosts remain in our genome, free from the pressure of natural selection to slowly rust.* Having established in 1953 that DNA was a cork-screwed ladder, and that it had the twin powers of replication and coded meaning, the biggest challenge in biology after Crick and Watson's great leap forward was to decode DNA, and this meant first identifying which bits of DNA were genes and which were not.

Imagine that every sentence in this book were a single gene: the human genome would be a book forty times longer than this one, filled with random text but with my sentences distrib-uted randomly throughout it. How would you identify which of them were the relevant sentences? Language is studded with punctuation to complement the letters and add composite meaning above the words themselves. Indeeditisprettydifficult forustoextractmeaningfromasentencethathasnospacesorpunctua tioninit. Fortunately for science, and necessarily for the cell, DNA is no different. Before scientists could work out what the words

---

* When this was first known in the 1960s, the term 'junk DNA' was introduced to indicate its redundancy. But it's now clear that much of it is far from rubbish; it is merely DNA that is not genes, and currently unexplored.

and sentences of DNA mean, the first challenge was to work out where the spaces were – where a gene begins and ends – and this meant working out its punctuation.

By the 1960s, scientists knew that life was built of or by proteins, that proteins were built from amino acids, and that DNA was the hereditary matter that coded the proteins. The big gap was getting from one to the other, from DNA code to protein. By experimentally inserting a molecule in between two letters of functioning DNA, Crick and another future Nobel laureate called Sydney Brenner intentionally disrupted its code and prevented a protein from being produced. These disruptions are known as 'frame-shifts mutations', like a film projector whose shutter speed is wrong, so you see half of one frame and half of another. By interrupting DNA with inserts equivalent to one, two or four bases, the same effect was observed – a disrupted protein. But with an insertion of three, a protein product was still viable. They deduced from this that the code of DNA works in sequences of three bases. This pattern is known as a 'reading frame': it places spaces in a length of DNA, exposing the meaningful triplets of letters.

Starting there, an American and a German scientist cracked the first coded message in DNA in 1961 using a process of elimination. Instead of trying to work out how a naturally occurring sequence of DNA translated into a protein, Marshall Nirenberg and Heinrich Matthaei strung together a length of genetic code that only consisted of the base thymine (the letter T). They inserted this into the mechanics of a working cell and provided it with a ready supply of amino acids from which a protein might be built. They knew that there are only twenty amino acids from which all life is built, so in each of twenty different test tubes containing all of them, a single different amino acid within the mixture was tagged with a radioactive label. This meant that if a protein that resulted from their doctored DNA buzzed with radiation they would be able to identify which amino acid was coded by a triplet of thymine bases. When they extracted the proteins from their various test tubes, the result was nineteen duds and one that set the Geiger counters buzzing. They discovered that a genetic code consisting exclusively of the letter T resulted in a protein consisting exclusively of the amino acid phenylanine.

And so the nature of the irrepressible code was known. Over the next few years, by varying the template, the unique triplets encoding the other nineteen amino acids were discovered, until by the end of the 1960s we had a complete readout of how DNA encodes proteins.

So here is a small section of DNA, part of a gene:

cctgggaccaacttcgcgaagcgggaagcccggcgg

Here is the same sequence broken into the triplet reading frame, as the cell reads it:

cct ggg acc aac ttc gcg aag cgg gaa gcc cgg cgg

And here it is again with each amino acid (here written in their abbreviated form) alongside each codon:

cct ggg acc aac ttc gcg aag cgg gaa gcc cgg cgg
Pro Gly Thr Asn Phe Ala Lys Pro Glu Ala Arg Arg

That string of amino acids forms part of a protein.

So how could scientists tell where a gene begins and ends? There is also punctuation in the language of DNA. In the continuous run of As, Ts, Gs and Cs, the cell knows where a gene begins because they all, without exception, start with the letters ATG, the so-called 'start' codon, like a capital letter at the beginning of a sentence. Similarly all genes end with a full stop, a stop codon, of which there are three: TGA, TAG and TAA. A reading frame for an entire gene always begins with ATG and ends with one of these three stop codons.

Proteins are, then, long strings of amino acids as decreed by the DNA that encodes them. They perform their functions by folding up into three-dimensional shapes, and the grooves and holes and clamps and pockets in their folded shapes give them all manner of abilities.* Proteins also team up to gain new purposes. For example, haemoglobin, which carries oxygen around your body in red blood cells, is made up of four proteins, together carrying a single atom of iron. The astonishing properties of spider silk are the result of the sophisticated complex of different proteins of which it is made, some of them neatly folded, others overlapping to create a high-tensile strength comparable with that of steel. Some proteins are enzymes, which catalyse bodily reactions, the metabolism in cells that keeps us alive. Others are sensory, like the ones embedded in the rods and cones of your retina, so specialized that they can detect a single photon of light and trigger the process of vision. All these properties are the result of proteins folding, connecting and interacting with others in highly precise ways. Diverse though the actions of proteins may be, the code underlying them is the same in all.

This is the bedrock of biology: the translation of code into action. But what is the process by which this translation takes place? If the genome is a sort of central office containing the plans to the factory, the plans never actually leave the office, so the relevant pages have to be photocopied. In other words there is an intermediary between the DNA and the site of the manufacture of the protein, and this envoy comes in the form of DNA's cousin, RNA.

DNA is a helix twisted from two struts of a ladder, the rungs linking each together. But RNA is a single strand, the rungs exposed. When a particular protein is required, the double helix of DNA splits in two to expose the relevant gene. A single strand of RNA is laid down on top of the exposed gene and each letter of that gene is copied into the RNA in mirror form. This envoy is called, appropriately enough, 'messenger RNA', and it carries the message from the genome to the site of protein construction.

---

* That sequence, from the retinal gene Chx10 above, folds into the shape of a grip that fixes onto short unique stretches of DNA and instructs the cell to activate another gene, rather like using a word in the index to seek out a particular sentence.

Later on we will see that RNA plays a far more central role in the origin of life than its mere messenger status suggests. However, we have now arrived at the first and biggest clue to the question posed by the phrase 'into a few forms or into one'. The system described above is truly universal. There are no life forms we know of that do not employ and entirely depend upon it: DNA – made of four letters – translates into proteins – made of twenty amino acids. It is known as the 'central dogma': DNA makes RNA makes protein. The fact that all known life is utterly dependent on this system makes it seem almost inconceivable that it is not related by a single, common origin. Certainly, working out how such a system might come about will be essential if we're to understand how life as we know it came to be.*

The Right Hand of Life

If the shared code and tools of life are not enough to point un-equivocally to a single origin of life, here's an oddity of biology that emphatically seals the deal. Hold your hands up with your palms towards you. They are reflections of each other. If you slide one in front of the other, palms still facing towards you, one cannot obscure the other as your thumbs stick out. This, of course, is the basis of the glove industry: a left glove won't fit on a right hand. And so it is with the molecules of life. In the same way that there are mirror forms of hands, so too are there mirror forms of certain molecules.

An atom is made up of a central nucleus, which is positively

---

* In a world that began to reject dogma during the Enlightenment, it is perhaps a pity that this pivotal process for all living things should have been named thus. It was Francis Crick who called it 'dogma', and years later he expressed regret that his intended meaning with this word was perhaps not the same as everyone else's. Over the years, as with all rules and laws in science, it has been refined and modified, so although it remains utterly true, it's not quite as dogmatic as its moniker suggests. All the same, that's what it's called.

charged, surrounded by negatively charged electrons. These electrons form the bonds that hold individual atoms together in molecules and, like the poles on a magnet, these electrons repel each other. So they attempt to maximize the space between them. When atoms bond with other atoms to make a molecule, they will space those bonds as equally apart as they can, and so tend to adopt, wherever possible, symmetrical forms. An atom that has three available bonds, such as nitrogen, will tend to form a triangular molecule, with a bond at each of its corners. Carbon dioxide, meanwhile, is a straight molecule: the carbon atom has four available bonds, oxygen has two, and so the carbon atom is flanked by two oxygen atoms, each occupying a double bond. But if an atom naturally seeks to make four bonds, as carbon does, the furthest spacing of bonds is not in two dimensions but in three: a pyramid with a triangular base. The carbon atom sits in the centre of this shape, with the four points equally spaced from each other. That's all well and good if all of the atoms on the corners are the same, or even if two or three of them are different from each other. But as soon as all four corners are occupied by a different atom, handedness becomes a possibility. In other words, exactly the same atoms in the same conformation can be arranged as mirror forms. In chemistry, this is called 'chirality', from the ancient Greek word meaning 'hand'.

We sometimes describe life on Earth as being 'carbon-based', by which we mean that all DNA and proteins are built from frames featuring carbon atoms. And carbon, as we have seen, can form 'handed' molecules, mirror images of each other, so it might be reasonable to suppose that we would find a mixture of right- and left-handed carbon-based molecules in life. What is remarkable is that, wherever life is concerned, the amino acids that make up proteins are left-handed.*

---

* The naming conventions that determine whether a chiral molecule is left or right are murky and not very helpful. In general, naturally occurring amino acids are referred to as L-amino acids, and they make left-handed proteins.

It was the killer of that persistently unfruitful idea of spontaneous generation who first identified chemical handedness. One of
the ways we can spot the chiral nature of a molecule is by simply
shining light through it.

Imagine tying a slinky to a wall and then waving it in all directions. The waves would be up and down as well as side to side. If
you then fed the slinky through a vertical slit and did the same,
only vertical waves would pass through. Certain molecules have
exactly this filtering effect on light. They can polarize light into a
single plane despite its natural tendency to wave in all directions.
Louis Pasteur noticed that shining a beam through a solution of a
simple molecule called tartaric acid that had been purified from
wine finings caused the light to be polarized.*

But he also noticed that tartaric acid that had been made in the
lab did not have this property, despite being chemically identical.
The reason is this: the molecules in tartaric acid are based on carbon, and when synthesizing the molecule in the lab, both left and
right-handed version are made in equal measure, the random
chance of flipping a coin. This means that any polarizing effect of
one version is cancelled out by the opposite action of the mirror
molecule. But the tartaric acid found in finings was naturally synthesized in yeast cells as part of the wine-making process. As a
result they are all right-handed. These partisan molecules would
only allow one plane of light through.

This minor diversion into the realms of wine and chemical
bonds and spatial thinking is absolutely essential, because, for reasons we don't fully understand, proteins only use left-handed
amino acids. It is a mystery why all of the twenty amino acids that
all proteins are made of are left-handed. Around four-fifths of

---

* Pasteur was French, and therefore, I imagine, was probably jolly interested in
wine. In fact, he developed the technique that bears his name – pasteurization –
to sterilize not just milk but wine as well.

humans are right-handed, so the equivalent with people would be as if southpaws simply did not exist.

Pasteur's revelation that the non-biological production of handed molecules resulted in left- and right-handed versions in equal measure was foretelling of one of the most dreadful medical tragedies in history. The drug thalidomide is a mild but effective tranquilliser and painkiller. Its efficacy across a range of conditions from insomnia to headaches earned it the status of a 'wonder drug'. But it was also an effective treatment for preventing vomiting, and so was offered to pregnant women suffering from morning sickness. Between 1957, when it was introduced, and 1962, when it was globally withdrawn, more than 10,000 babies were born with severe birth defects including stunted limbs and other abnormalities as a result of being exposed to thalidomide in the womb. Thalidomide is a chiral molecule, and it was determined that, of the two mirror versions, only one had the mutating effect. Like in Pasteur's tartaric acid, its production in pharmaceutical factories did not account for the two mirror versions, and the drug was sold with both hands in equal measure. We now know that either version of thalidomide has the confounding ability to flip to its reflection once in the body, so even in its harmless version, it would potentially be harmful to the developing embryo. It's still prescribed in some countries as an effective treatment for leprosy and other illnesses. But thalidomide's use by pregnant women is necessarily and strictly forbidden.

This handedness is a biological phenomenon, with proteins being almost exclusively on the left. The origin of this prejudice is not known. It might be as simple as a chance event, but one that stuck. The uniformity points towards the development of a system with a single point of origin. If life had evolved twice, the likelihood would be that we would see left- and right-handed proteins.

DNA has a similar but opposite steadfast bias: it is always right-handed. Point the index finger on your right hand, and draw an imaginary clockwise circle in the air. At the same time move your hand away from your body. That is the turn of the double helix in its most common form, like a typical wood screw.* Even though that mirror-image double helix could perfectly well exist, it just doesn't. Life, yet again betraying its unique origin, only uses right-handed DNA.

The fact that a mirror world could exist but doesn't shows forcefully that life is of singular origin. If there ever was a choice between left and right, only one was selected, and the other discarded, never to feature again in the natural course of things. At the beginning of these biological mechanics, it might be as simple as a flipped coin, a chance event that was followed with absolute fidelity forever.

The Singularity

The origin of cells from parent cells and the origin of species via slowly changing genes in those cells both bear the hallmarks of a single origin. Those three aspects of biology – cells only from existing cells, DNA changing through imperfect copying, and modified descent of a species as a result – logically unveil a single line of ancestry that inevitably leads back to a single point in our deep, deep past.

In other words, any one of the new cells that were pulled from their parent as a result of the orchestrated disaster response to a paper cut bears a direct and noble lineage. Life is replete with

---

*There is a website dedicated to outing depictions of DNA that spiral the wrong way, the Left-Handed DNA Hall of Shame, to which I am sorry to say I qualify alongside most of the major science journals and websites, not to mention newspapers, and many, many adverts and films. There is one organization whose logo features the alien leftie screw, and that is the Astrobiology Society of Britain. These are the researchers whose work concerns the existence of life in the rest of universe. Whether that was a design error or a deliberate play on the fact that they are primarily concerned with unearthly biology is unknown. As far as I'm concerned, it's quite clever.

extinction, but a new cell on your hand is one of the great survivors of the longest pedigree in history. In one generation its ancestry traces back to the fertilized egg from which every cell in your body was born. The egg and sperm which fused to form your unique combination of genes and DNA can trace a similar pathway back to their fertilized egg origin, and so on through their parents, grandparents and every ancestor in your family tree, and indeed the history of our species.

Of course, it doesn't nearly end there. These cells in turn trace their ancestry through the lifespan of our ape-like ancestors. We don't know exactly who they were yet, but they existed in the bodies of individuals who first carved bones to make tools, who flint-started fires, and stood upright as we, amongst apes, uniquely do.*

And it certainly doesn't end there either. Before being in the loins of that upright ape, your cell lineage was carried through long-gone primates, maybe something like *Proconsul*, who resembled something between a chimpanzee and a macaque. Before them, it lay in monkeys and earlier through their furry, wet-nosed ancestors, more like modern lemurs. And as we follow them back, at some point, cells that were once the predecessors of your Pac-Man macrophages or sparking neurons existed in a shrew-like creature with fur and nipples. That critter, your ancestor, was present when an enormous meteor thundered into the Tropic of Cancer and called time on the reign of the dinosaurs sixty-five million years ago. Your cell lineage witnessed the entire fall and rise of those beasts inside many forms of early mammals, such as the *Cynodontia*, which looked like a snarling rodent of unusual size. Before those first mammals more than 220 million years ago, your cell's ancestor was a shelled egg of a reptile, hairless and

* It's not just the fossilized bones that tell us our past. Trace fossils are also crucial. The upright footprints set in soft ash 3.4 million years ago in Laetoli in Tanzania by *Australopithecus afarensis* are stunning evidence of our unique gait, but the truth is that we don't really know if these ape-like people were our direct ancestors. The fossil record for human evolution is desperately uneven.

cold-blooded, such as *Diadectes*, a two-metre-long brute that looked like a stocky, diet-shy crocodile.

To get to be in that egg, it survived iterations of beasts that had, amongst many other things, evolved cells that produce a thicker collagen framework in their breathing organs; with this novelty the primitive lungs could support themselves without the need for water. Your cell line may well have passed through a creature like the 375-million-year-old lungfish *Rhinodipterus*, which, unlike other fish at that time, had a muscular neck, and with that the ability to raise its head above the water line. These innovations allowed it to breathe air rather than suck oxygen from the seas, and maybe opened up a whole new world of food. Back further your cell line was contained within a much more fishy thing, with fins and gills. Before that it was in one of the first vertebrates, a swimming thing that looked a bit like an eel or a lamprey. And before that it was in a much wormier thing, similar to the living two-inch amphioxus: in one of those cells a massive copying error in its genome, a quadrupling of its entire DNA, resulted not in a fatality, but in a whole genetic platform on which vertebrates began to evolve. Before that it was in a spongy thing. And before that just a clump of cells free floating or lodged on a rock. And back and back and back ceaselessly into the past. The lineage of your cells has survived every disaster, catastrophe, meteorite, every extinction, ice age and ravenous predator, every event in this solar system for almost four billion years.

The vast majority of your cells, including all of the new ones in your cut, are terminal branches in this absolute pedigree, as their story will end with your life. The only ones that survive to give rise to cells in the next generation are sperm or egg. And of the trillions of cells that have worked for you during your life, only a handful will go on, the lucky few that will meet a sperm or egg and make a child. But the information within all of them will be carried on. Down that line has been transmitted the DNA that builds all of those cells together to make the most efficient way of ensuring the perpetual existence of your genes: an organism. Life is an astonishingly conservative system: DNA is the same in all species; the letters of the code are all the same; the encryption in the code is the same; even the orientation of the molecules is the same. What's true in a bacteria is true in a blue whale. Only a system with a single root could display such conservation.

That pathway, retracing our steps as they become fainter and fainter over geologic time, can be applied to any creature alive today, or ever. The gaps become bigger, and it's almost always only ever hypothetical. Although we have a good overall understanding of the origin of species, to claim one species was the direct ancestor of another is often overstating what we can know. But the broad sweep of evolution is well understood, and any step backwards through the past from any creature drives us to a single conceptual home. The branching tree of life ultimately becomes narrower as we reverse through time until we reach a single stem.

We could trace an equivalent route backwards for a cell taken from the boiling mud in an Icelandic hot spring, from the flower of a sweet pea, or a button mushroom from the supermarket, and every time, we would end up in the same place. In every cell is a perfect unbroken chain that stretches inevitably back to the origin of life. That lineage irresistibly leads to one single entity, which we call the 'Last Universal Common Ancestor', or Luca. Somewhere on the infant Earth, Luca split in two. Since then, the thing that we struggle to define as life has passed uninterrupted from it to you, via a colossal series of iterations. Existence is bewilderingly tenacious.

Entities that we might be able to describe as living may have emerged several times, but life only enduringly survived once, and then continuously. We are confident of this because these other forms simply do not exist. At least they have not been discovered: there is a branch of speculation that proposes the rather futuristically named 'Shadow Biosphere'. This is the idea that there is a second (or more) undetected tree of life on Earth, with hallmarks different from the ones on the only tree of life we know of. But as it is, every life form so far examined is based on cells, DNA and Darwin.

Discovery of a second tree of life here on Earth would give much-needed credence to the search for life on other planets, as it would double the number of known successful origin of life events. It would show that we are not a fluke. However, science is based on observable evidence. Therefore the Shadow Biosphere, whilst sounding quite thrilling, is resolutely science fiction.

Pruning the Tree of Life

The next questions seem glaringly obvious. What was Luca, and where did it come from? We can reasonably assume that it had DNA as its genetic code, as all creatures after Luca do, and they are unlikely to have evolved that mechanism independently.

These ideas are largely based on what we share with other lives, whether it is the orientation of molecules, or big physical characteristics like five-digit limbs, or even more simple things like having a head at one end and a tail at the other. Since the era of DNA began, and more so when the technology for reading genomes became much more accessible in the 1990s, studying evolution was fortified by comparing similarities and differences in the precise letters of DNA. Because we have shared ancestry, thousands of our genes are very similar in closely, or even distantly, related species.*

DNA acquires copying errors at a fairly steady rate, which means that we can compare DNA in any living species, and work out when they separated. We can compare genes and protein

* A gene called Pax6 is a beautiful example, and one on which I once worked. In humans, this gene encodes a protein that has the role of dictating that an area of the developing brain will mature in eyes. It does the same job in mice and zebrafish, whose versions of Pax6 only differ by four out of a hundred amino acids, despite having a common ancestor around 400 million years ago. The conservation is such that in the fruit fly beloved of genetics researchers Pax6 also specifies where an eye is going to grow, even though it is a fly eye, not a mammalian one.

sequences from any two organisms, and calculate how long they have been diverging. In this way, we can reconstruct history in exactly the same way as palaeontologists do with fossil bones, by looking at similarities and differences, and assembling all those comparisons to show not just relatedness between two species, but when the schism occurred. This is called 'phylogenetics' and has utterly confirmed Darwin's ideas about the branching tree of life.

However, biology is littered with exceptions of varying sizes, and when it comes to the Tree of Life, there is one huge exception. Using phylogenetics, many scientists now argue persuasively that for the first billion years or so life was not so much of a branching tree but a tangled bush.

The first life forms, for the first couple of billion years following their split from Luca, were single cells. They were evolving, but not really transforming radically. In fact, despite a generational time orders of magnitude speedier than most animals, for the first couple of billion years life failed to make it past the stage of microbes. These two domains of life are archaea and bacteria – superficially similar things, both single-celled entities, and roughly the same size. For a long time, archaea were similar enough not to be recognized at all. But they are different enough to now be classed as very distinct from bacteria and indeed everything else (and we will find that these differences are crucial in the theory of the origin of life). As a domain, they are a separate category at the highest rank of how we categorize life. The great leap forward occurred with the arrival of complex life. This branch of the tree, the third domain, is called eukaryotes, and includes everything that isn't in the first two, including you and me and yeast and snakes and algae and fungus, flowers, trees and turnips. At some point, maybe some two billion years ago, complex life emerged when an extremely unlikely pairing occurred: an archaea swallowed a bacteria. Rather than death of one or both, mutual benefit was the result. The consumed ceased being a free living entity and was permanently annexed into the guts of what would grow to become the third domain of life, the one that you are in.

This idea was first presented by the American biologist Lynn Margulis in 1966. It caused rousing controversy, was largely rejected and she was considered to be heretical. With time, experiment and evidence, her views were vindicated and are now the orthodoxy. The evidence comes in a number of forms, most simply that complex cells, from which we and all animals are made, contain small internal power units called mitochondria. The processes that happen inside these power stations are of fundamental importance to the origin of life, but we will come to that in due course. These function as chemical engines that provide energy for the cell and, by extension, the organism. In simple terms, mitochondria resemble bacteria. They are roughly the same size as bacteria and just like bacteria have circular strings of DNA as their own genome, which are independent of the vast majority of a cell's genetic information, safely housed in the nucleus.

That engulfing of one into the other was not a meal, but a hostile takeover. The engulfed would never be free again, but enabled previously impossible growth in its new host. It came with its own genome, with thousands of genes. Over time, most have been lost to natural selection, or migrated to the host's nucleus control centre. But the mitochondria retains an independent set of genes to this day, almost all of which are devoted to maintaining the energy generation for cells. When this happened, charged with new energetic vigour, genomes could grow and form greater templates for evolution beyond single cells. Cells could evolve internal structures and compartments that increased specialization of cells. From there, the acquisition of coordinated communication between cells meant that an organism was not restricted to a single cell. Multicellularity followed, eventually enabling the evolution of body plans for plants and animals, complex networks of crosstalking cells that interact with each other and the environment in harmony.

For back-tracking through the tree of life in order to work our way back to Luca, these events provide a problem. The complication in estimating the timing and a description at the base of life comes from the fact that both bacteria and archaea do something else scientists did not expect. We pass genes down only from cell to daughter cell, from parent to child. Bacteria and archaea can swap genes, and therefore characteristics, with other individuals. Sometimes they don't even need to be of the same species to do this. This is called 'horizontal gene transfer' (as opposed to 'vertical' descent), and it is critical in confounding our attempts to understand the origin of life. The reason for this is because these cells don't always gain evolved functions by the typical process of descent via cell division.★

The evolution of language is a handy metaphor here. The words bigamy, bicycle and biscuit have a common root. 'Bis' in Latin means 'twice', so you get twice married, ride on two wheels and eat twice-cooked tasty snacks. But there are some words or phrases that are

---

★ It is precisely this behaviour that has enabled so-called superbugs to become such an infectious problem in hospitals. Within a few years of the introduction of penicillin in the 1940s, a previously undetected strain of the bacteria *Staphylococcus aureus* was described that was resistant to penicillin. What this means is that in the presence of the otherwise lethal antibiotic, a random mutation appeared in one bacteria that rendered penicillin ineffective. As bacteria reproduce with alarming speed, soon this penicillin-immune microbe had spread. So, we invented methicillin, another antibiotic that would still prove lethal to the penicillin-resistant ones. Guess what? Within a few years, Methicillin-Resistant *Staphylococcus aureus* were found, MRSA. Evolution has provided the mechanism for survival, and bacteria use it with tenacious vigour. Once a bacteria has randomly developed resistance to an antibiotic, it is more than happy to share it not just with its descendants, but with its neighbours. They extend a thin bridge, a *pilus*, which brings two bacteria together, and through a tiny pore sections of DNA can be transferred from donor to recipient. This vastly speeds up the spread of an advantageous trait. It means that a population doesn't necessarily have to wait for a random resistant mutation to arise and to be passed from parent to offspring. It might be that the resistance is already in place, unused, and is deployed throughout the population as soon as they are exposed to the antibiotic. It's a familiar story: we try to control an organism, it evolves to survive. Humans taking on bacteria in warfare is a bold gesture. Here, and elsewhere, the words of the great chemist Leslie Orgel resound: his so-called second law states that 'evolution is cleverer than you are'.

better expressed in other languages or even absent in the recipient, and so just get pinched. *Cul de sac* has a slightly different and more specific meaning in English than 'dead end', so we have adopted it. The German word *Schadenfreude* has no English counterpart, but is now a splendid word in itself for ignobly enjoying your enemy's misfortune. But in Swedish, the word is *skadeglädje,* derived and adapted directly from *Schadenfreude* but bypassing any common root. It has passed horizontally, and subsequently evolved independently.

What this horizontal gene swapping means is that our techniques for tracking back through the history of life using DNA as our guide may only work convincingly from the point where descent becomes vertical. The tree of life only begins to resemble a branching thing that deserves the name 'tree' *after* the emergence of complex life. From that species on, when an archaea swallowed a bacteria, the overwhelming majority of inheritance was from parent to offspring: descent with modification. UCL biochemist Nick Lane calls this the 'genetic event horizon': comparing genes will take us all the way back to the point where it looks like a nice branching tree, but our vision is blurred before that point. It just becomes far too messy to unpick the deepest past.

So, while we can sensibly use DNA to infer that humans and chimps had a common ancestor around six or seven million years ago and humans and the meagre sea worm amphioxus had a common ancestor around 500 million years ago, we cannot reliably use this to date Luca. The branches of the tree become tangled and scrambled before complex life, and tracing the changing patterns of DNA as species evolve is impossible. We are left, then, with very little genetic evidence of what Luca actually was at all.

The Creation of Luca

We can figure out some things about Luca, though. Calculations and logic predict that the last universal common ancestor had characteristics that are shared between archaea and bacteria, and that means genes, proteins and the cellular mechanics largely similar to what we see today. That means that we can use comparisons of these molecules to understand some things about Luca, even if we cannot apply an accurate date. A study by Douglas Theobald at Brandeis University, Massachusetts in 2010 applied hard statistical analysis to the domains of bacteria, archaea and complex life. He looked carefully at the construction of twenty-three proteins that are present in each of these domains with seemingly common descent, like words that sound and mean similar things in different languages. Based on the similarities of the sequence of amino acids that make up these proteins, Theobald calculated that the odds of their having arisen independently was 1 in $10^{2,860}$ (that is, a 1 with 2,860 zeroes after it).*

Another clue to Luca's singular origin concerns the most fundamental cellular machinery – the ribosome. It exists in all cells as a processing plant organized in minuscule blocks of molecules. Its role is universal, and as such again supports the single origin idea. The ribosome reads the genetic code and translates it into protein. The intricacies of this exquisite machine are explored later in the book (and particularly in the Afterword to the other part, p. 93), but in essence its job is to read the genetic code (already transcribed into an RNA version), and translate each three-letter section into an amino acid. The ribosome strings together the amino acids in accordance with the messenger RNA, and a protein feeds out like a ticker tape.

We look to the ribosome as a useful indicator of relatedness because it is fundamental – without its manufactory, we have no proteins, and none of the life we are aware of can exist. We can tell that mitochondria in complex cells are derived from annexed bacteria because mitochondria have their own ribosomes, separate

* And don't think that this has settled the matter. A Japanese team has published a response that says this analysis is not sufficient to reject the idea of multiple origins. That's not to say they or other critics are advocating a multiple-origin hypothesis, just that this particular study doesn't prove its own conclusions. Scientists can be a quarrelsome lot.

from their host cell, and these ribosomes are much more similar to bacterial ones than animal ones. We can compare the sequences of the genes that encode parts of the ribosome in all species, and trace backwards the changes that have occurred over time to predict what Luca's ribosome looked like.

This is a fruitful endeavour, but the results are contested. For example, by looking at the particular sequence of ribosome parts across many species, you can reasonably infer the temperature at which that host organism thrives. Several parts of the ribosome are built from neatly folded RNA molecules, themselves made from letters of genetic code A, C, G and U. As in the double helix of DNA, C pairs with G and A pairs with U. But C and G form a bond that is more stable at higher temperatures. So, we can infer from the relative amount of CG bonds in ribosomes a preference for warmer conditions. This is borne out across many species, and the ribosomes in which we see the highest CG content are from high-temperature extremophiles – in other words, organisms that thrive in heat. Where do we find such creatures? In many places, but the most impressive habitat is in and around hydrothermal vents under the sea. There, heated by a rip in the Earth's surface, plumes of noxious chemicals chug out which can boil the sea. But life still abounds there. Dozens of species thrive there: bacteria, archaea and even large complex creatures such as the Pompeii worm, which can endure temperatures up to 80 degrees Celsius (not least because it wears an insulating fleece of hardy bacteria).

Some models of Luca's ribosomes suggest that the amount of Cs and Gs was disproportionately high in their component parts. This might suggest a hot home for the base of life. Certainly, the discovery of extreme heat-loving, or hyperthermophilic, archaea and bacteria in environments such as the hot springs in Yellowstone Park, or submarine thermal vents, supports this idea, as these organisms mostly occupy spaces at the base of evolutionary trees, as much as we can reconstruct them.

But the simple truth is we don't – possibly can't – know. Reconstructing the past using phylogenetics is a complicated art, with many confounding factors. Using the ever-changing DNA sequence, we can't see back to Luca because of bacteria and archaea's ability to move genes sideways, not just from parent cell to daughter. That's not to say that Luca did not have specific characteristics that we can investigate and compare with those in living things. It's just that the idea of a single cell, a biological equivalent of the Adam from the biblical Genesis, might be naive. If Luca was a cell, even in its most basic form, it still would have systems within it that are like their modern counterparts: DNA, RNA, proteins, ribosomes to make proteins, a cell membrane, and crucially, a highly developed way of capturing energy – a metabolism; if not, then we would expect to see different and divergent mechanisms in bacteria and archaea. This is not to denigrate this useful species. Luca's real use for science is as a proxy: if Luca is at the root of cellular life, it represents an all-dominant coagulation of what came before. Bill Martin, a brilliant and pugnacious origin-of-life biochemist who we will meet properly later, says that the trouble with Luca is that, 'like love, it means different things to different people'.

It's almost three and a half centuries since cells were bled, ejaculated and fished out of their natural environment and seen down a primitive microscope. Since then, we have picked them apart to the extent that we have earned almost full command of their facilities, acquired over four billion years. We see their commonality so clearly, a beautiful neatness where everything we discover in biology serves to refine and reinforce the truth of evolution. It's a wonderful state of affairs, and reflects the maturation of a science. The essential qualities of life are known and gel into a grand vision: life shares its tools, its processes and its language. The security of a robust unifying theory of life as we know it allows us to quiz a much more difficult puzzle: where did Luca come from? As it happens, the best way to begin to understand the emergence of life on Earth is to take a hard look at where and when it happened. Therefore, we must start at the very beginning. It's a very good place to start.

3.   Hell on Earth

'Long is the way and hard that out of hell leads up to light.'

John Milton, *Paradise Lost*

If you want to construct a picture of the Earth on which life first emerged, think about how we've named it. There are four geological eons spanning the Earth's 4,540-million-year existence. The most recent three names reflect our planet's propensity for living things, all referring to stages of life. The second is called the 'Archaean', which rather confusingly translates as 'origins'. The third is the Proterozoic, roughly translating from Greek as 'earlier life'; the current, the Phanerozoic, starting around 542 million years ago, meaning 'visible life'.

But the first, the period from the formation of the Earth up to 3.8 billion years ago, is called the 'Hadean', derived from Hades, the ancient Greek version of hell.

Life does not merely inhabit this planet, it has shaped it and is part of it. Not just in the current era of man-made climate change, but all through life's history on Earth, it has affected the rocks below our feet and the sky above us. And necessarily the origin of life is inseparable from the fury of the formation of the Earth in the first place. A picture of the Hadean Earth is key to understanding the wild natural laboratory in which life contrived to be born. Just as the formation of our home world is an event in space, we'll come to see how the emergence of life here is essentially a cosmic event.

The study of the early geology is all rock-hard science, but the evidence is often both literally and metaphorically thin on the ground. It requires geological detective work with clues dotted around the planet, and off-world too. Geology gives us hints to how the Earth formed from the bits and pieces of matter floating in space around the Sun. But it also begins to describe the world that will evolve into the host of the only living things that we know of. Whilst our lives are built on the stability of the Earth, we also are keenly aware that our planet is sporadically violently active. The solid surface (including the sea floor) is made up of seven or eight leviathan continental plates, and a collection of smaller ones. These all float on the slowly flowing but solid rock of the mantle, itself encapsulating the molten core. The plates that form the crust are in constant flux, unhurriedly jiggling together. Some, such as the Pacific and North American plates, grind against each other, forcing up new land inch by inch. The subcontinent that is now India was once an island, and crunched into mainland Asia in a process that began around seventy million years ago, inching forward and crumpling up the land into the mountains of the Himalayas. These mountains will continue to grow at a rate of a few millimetres every year as the Indo-Australian plate continues to muscle its way into mainland Asia. Other plates are pulling the Earth apart at the seams. The United States' colonial expansion continues westward every day, as the coast of Hawaii grows new land at a rate of a few metres per year with molten rock gurgling up above sea-level and solidifying. Earthquakes shake the land, and the seabed, dislodging mountainous blocks of water that become tsunamis such as the one that wrecked the east coast of Japan in 2011. Today, these events are anomalies, though they remind us that our planet is alive not just with cellular life but also with slowly flowing rock. But for the most part our Earth is reassuringly stable.

Not so in the past. The birth of planets is a process of summoning order from the chaos of the early solar system. Violence ensues. The Sun, the star at the centre of our planetary system, formed around 4.6 billion years ago as a colossal cloud of free-floating molecules collapsed under its own gravity and condensed into the huge nuclear fusion reactor that continues to heat the Earth today. In the immediate aftermath, the Sun sat in the centre of a solar nebula, a flat disk of detritus left over by its formation but held there by its own gravity. Over the course of the next few million years this matter, mostly dust and gas, began to stick together in clumps. At first these were the size of medium-sized concert halls, but over the hundreds of centuries, lumps collided and stuck together in a process called 'accretion'. Closer to the hot Sun, the temperature is higher, which makes it harder for gases to condense and accrete. It's for this reason that the inner four planets of the solar system – Mercury, Venus, Earth and Mars – are terrestrial, made of rock, whereas the outer four – Jupiter, Saturn, Uranus and Neptune – are gaseous (or frozen).*

It's impossible to describe a whole planet in simple terms, as we know very well how our home is not a unified land mass. Much of the surface of the modern Earth is solid, more is ocean. Of the grounded parts, there are extremes of temperature and geography, snow, deserts, marsh, forests, plains, mountains and so on. The rock beneath your feet is likely to be very different from the rock underneath a reader's feet on the other side of the planet. Similarly it's virtually impossible to describe simply what the Earth looked like in the Hadean. The evidence is vanishingly sparse, and as a result this period unhelpfully gets called the Cryptic Era. But we can extract meaningful models in broad terms. We know that immediately after accretion, the planet was probably largely molten. But, contrary to previous thinking, we now think this period of molten Earth was short-lived. There are no surviving rocks from that time, so in the past we had assumed that there were no rocks. But the absence of evidence is not the same as evidence of absence.

Just as we look in rocks for the traces of recent life – the fossil-

* Pluto, once the ninth, is no longer considered a fully fledged planet, as it is one of several similar-sized bodies in that area of the solar system.

ized forms of dinosaurs, or even cells, for example – our planet's history before life is embedded in geology. The oldest dated matter on Earth comes not from rocks, but in the form of zircon crystals, an abundant mineral found all over the world, probably best known as a cheap substitute for diamonds in jewellery.

Zircons have two convenient characteristics. The first is that they can withstand metamorphosis – the brutal churning of rock over very long periods of time. The second is that the atoms of zircons are naturally arranged in a neat cubic structure. This molecular box can trap atoms of uranium inside them, as few as ten parts per million. A small proportion of uranium, like many elements, is radioactive, and over time will decompose into lead. Due to the precise nature of their crystal structure, when zircons form they can include uranium, but exclude lead atoms. Once imprisoned in this cage radioactive uranium slowly mutates into lead over the course of millions of years, and because this decomposition happens at a fixed rate (what we call a half-life), the moment a zircon crystal forms, it sets a clock at zero. Lead found inside zircons must have begun its life as imprisoned uranium, so by quantifying it we can date the origin of that crystal with 99 per cent accuracy. Out in Jack Hills in Western Australia, zircon crystals have been found that trapped their uranium 4.404 billion years ago.

Dating is not all we can tell from the constituents of cheap jewellery. We can also infer from those same crystals that their formation was part of a solidifying process, the development of a crust. This means that, although there are no rocks from that period, we can know that there was land in the Hadean. We can also determine what other ingredients were present. The Jack Hills zircons also harbour a particular type of radioactive oxygen whose presence looks like that found in modern crystals that were formed as the Earth's crust gets sucked down beneath the ocean floor. The presence of this oxygen suggests that from as little as 100 million years after the initial moulding of the Earth water was present. On

Earth, there is no life without it, though this water was likely to have been extremely acidic.★

So, in the early cryptic days on Earth, it might not have been quite the hellish inferno of endless seas of molten lava. Within a mere 100 million years, the Earth had a solid surface and oceans. Sounds pleasant enough, but let's not paint such a picturesque portrait. The earlier theories of a molten Hadean Earth were based on a robust observation: we can't find any rocks from that era. If there was a solid surface, what in hell's name happened to it?

It turned out that we were looking in the wrong place. In fact, we were looking on the wrong heavenly body altogether. To address the question of what happened to the early Earth, we sent twelve men to the Moon. The Moon itself was born out of the most destructive impact that the Earth has yet suffered. Somewhere between fifty and 100 million years after the formation of the solar system, the Earth had its worst day. It was struck by Theia – a terrifically inappropriately pretty name for such a harbinger of doom. Current theories suggest that Theia was a rock the size of Mars. It blasted enough matter from the embryonic Earth into space that it reformed as our closest celestial neighbour, the Moon. The impact was devastating, enough to rip the first atmosphere from the planet. Theia's glancing blow may be what shifted the Earth's axis from vertical to its off-kilter stance of 23.5°. This lean is what causes the seasons, as the distance from the Sun varies with the axial tilt of the Earth.

But it was what came after the formation of the Moon that concerns us. It's the characteristic pock-marked lunar visage that gave us clues to the state of the Hadean Earth. Between 1969 and 1972 Nasa's Apollo programme landed six missions and twelve

★The source of the Earth's water remains controversial. The absence of an atmosphere and its position in space so close to the Sun may have caused much water on Earth to evaporate. But it may be that the shape of the rocks that first accreted to form this planet held water in their crannies. Another idea that some scientists argue for is that most of the water on Earth was delivered in one or several icy comets whose frozen payload melted on delivery.

explorers on the Moon, beginning with Neil Armstrong's famous first small step. Over those missions, astronauts collected around half a tonne of rock and brought it back home for analysis. The last man to walk on the Moon, Commander Gene Cernan in Apollo 17, is quoted as saying that 'we went to explore the Moon, and in fact discovered the Earth'. There is a great truth in that quotation, as it was in the subsequent analysis of Moon rocks that we were to discover the nature of the Earth's formative years. Unlike Earth, the Moon has no atmosphere or winds, and no shifting geology, so the craters formed by meteorite impacts are left undisturbed, alongside the footprints of the Apollo pioneers. What that means is that we have a record of meteorite activity in the local solar system, footprints unscathed by the winds, seas and tectonic grind of the Earth. Lunar geologists dated rocks bearing the hallmarks of meteor strikes. These are called 'impact melt rocks', and they found that they all occurred in a precise window of time, between 4.1 and 3.8 billion years ago. We can deduce that this was a period of intense local meteoric activity, and by inference, that the Earth also suffered this hellish pummelling from above. The young solar system was crowded with debris and leftovers from its birth, and for a period of 300 million years until the end of the Hadean, we got the full brunt of it. This period is called the Late Heavy Bombardment, 'late' because it was mercifully the last time our home would suffer such a battering.

How heavy is heavy? Meteors fall from the skies all the time. Almost all, thank goodness, are tiny, and burn out then fade away in the atmosphere as shooting stars. Occasionally a big one hits, whereupon they become meteorites, such as one that fell on the small Australian town of Murchison in September 1969, just a few weeks after Apollo 11 returned Armstrong, Buzz Aldrin and command module pilot Michael Collins to Earth. That one weighed more than 100 kg, and carried a payload of interest to this story, as we will find out later.

If you're prone to making a wish when you see a shooting star, why not hope that we don't get to witness anything near the size of the best-known meteorite. Sixty-five million years ago a five-or six-mile-wide rock smashed into an area of what we now call Chicxulub in Mexico. The crater is now hidden, mostly beneath the sea, but its 110-mile-wide shadow remains, and was spotted by oil prospectors in the 1970s. In the ground and seabed, there is a broken but detectable ectopic circle of tiny glass beads that were forged from molten rock during the heat of impact. And from space we can see the same circle in minuscule gravity distortions only measurable from precision equipment in satellite orbit. There has not been an impact anywhere near that magnitude since then, and be thankful for that. The Chicxulub meteorite was the trigger that extinguished the reign of the dinosaurs, and paved the way for small mammals to evolve, eventually into us. An impact of that order means that it's very likely that the meteorite instantaneously wiped out many millions of creatures, with an expanding circumference of mile-high mega-tsunamis racing away from the impact site, levelling the land like a wave crashing on a beach. With that there also would have been a fireball hot enough to melt sand and rock into those telltale glass beads. But the full impact of the meteorite would have taken thousands of years, a dust cloud thrown up that blotted out the Sun. The Chicxulub impact irreversibly changed the Earth system, wiping out once dominant life forms. And yet compared to what was happening on the infant planet, the place on which life began, Chicxulub was a drop in the ocean.

Scientists have estimated that during the Late Heavy Bombardment something like fifteen astronomical rocks over 100 miles wide, twenty times the size of Chicxulub, bruised our world. Of these a handful were maybe 200 miles wide. For 300 million years, giant rocks rained down from the sky, some as big as decent-sized islands. The power of any one of the tens of thousands of impacts during this time would make the most destructive nuclear bomb seem like a firecracker. Global environmental destruction would have occurred at least every few centuries. Any potential surface habitat for living organisms would have been destroyed over and over and over again. The relentless pounding that the planet suffered during the Late Heavy Bombardment was enough to boil the oceans and vaporize the land.

And then, it significantly calmed down. The meteoric blitz of the Hadean ended around 3.8 billion years ago, leaving a frazzled Earth, still tempestuous and rough, but at least not barraged from the skies. The Sun was dimmer than today, probably less than three-quarters of its current strength. With that, the Earth cooled quickly, and water from volcanoes and comets condensed into oceans that covered the planet.

The precise point at which life began is unknown, and almost certainly unknowable. It may be that it began multiple times, maybe during the Hadean, but was wiped out all but once by the sterilizing spray of the Late Heavy Bombardment. One 2009 computer model by scientists in Colorado suggests that, even if the Hadean eon sterilized the surface of the Earth, life could have survived at the bottom of the ocean.

The general (though not unchallenged) consensus is that the first evidence for living matter dates to around 3.8 billion years ago, coinciding with the end of the Late Heavy Bombardment. These clues come in the form of that vitally important atom, carbon. Cells are not visible in the fossil record at this age, as rocks older than 3.5 billion years tend to have undergone the harsh geological metamorphosis which irretrievably churns up any shadow of living structures. So we have to look for the chemical signatures of life trapped in rocks. In a formation on the west coast of Greenland, rocks have been found that contain the merest trace of a form of radioactive carbon that has no earthly reason to be there, unless it had been processed by a living organism.

We don't know what that life form was: only by the presence of that carbon can we infer that an organism that had similar fundamental mechanisms to modern life existed all that time ago. Skip forward 400 million years and the remnants of life are abundant and much less controversial.* The best of these comes in the form of stromatolites: foot-wide stone mushrooms that sprout from the shallow seas in Australia and other locations around the world. They are formed when floating mats of solar-powered bacteria trap tiny particles of grit in their slimy mucus, and over millennia this floating scum slowly settles into layered lumps of stone.

Ingredients

But that is hundreds of millions of years of evolution after the end of the Late Heavy Bombardment. What we see from then are scant pieces bearing evidence of living things in a colossal jigsaw. We have a picture of the Earth, more settled than it had been for hundreds of millions of years, but still violent, with electrical storms, churning land masses, volcanoes chugging out gases into the atmosphere and tumultuous seas. This is a contemporary understanding of the early Earth, and has helped us formulate experiments and hypotheses for the conditions in which life emerged. However, the first speculations about life's emergence predate this by a century.

In 1871, Darwin wrote a letter to his friend Joseph Hooker contemplating the switch from inanimate chemistry to life. On the second page of this almost unreadable document† he considers not the origin of species, but the origin of life:

> It is often said that all the conditions for the first production of a living organism are now present, which could ever have been present. But if (and oh what a big if) we could conceive in some warm little

---

* I say 'less controversial', as some scientists contest that stromatolites could form from a non-biological process — abiogenesis. Nevertheless, the consensus leans heavily towards these rocks being strong evidence for an Archaean world thronging with microbial life.
† Darwin's handwriting was extraordinarily untidy, to say the least, and the original letter is barely legible.

pond with all sorts of ammonia and phosphoric salts, – light, heat, electricity &c. present, that a protein compound was chemically formed, ready to undergo still more complex changes, at the present day such matter wd be instantly devoured, or absorbed, which would not have been the case before living creatures were formed.

With this famous 'warm little pond' Darwin is prefiguring the concept of primordial soup ('primeval', meaning 'original' or the 'earliest time', and 'prebiotic', meaning 'before life', are also used, largely interchangeably). There, he lists the ingredients of the soup, just like a recipe. Though ignorant of our modern picture of the Archaean Earth, Darwin had casually meandered into what would become the dominant idea of the origin of life. He was not the only one to take these speculative baby steps. One of his biggest champions was the German zoologist and polymath Ernst Haeckel, an early proponent of the idea that biology and chemistry were continuations on the same spectrum. In 1892 he proposed a process in which occurred 'the origin of a most simple organic individual in an inorganic formative fluid, that is, in a fluid which contains the fundamental substances for the composition of the organism dissolved in simple and loose combinations'. Chemists were already dabbling in biological alchemy, not gold from base, but conjuring the molecules of biology from chemistry. In 1828, a German scientist, Friedrich Woehlerr, synthesized urea, a key biological molecule and a component of urine, noting in his methods that he had done it 'without the use of kidneys, either man or dog'. This contradicted the then-popular concept of 'vitalism', that life was somehow fundamentally different from non-life. Woehlerr had shown that the molecules of life could be made synthetically.

This idea that the birthplace of the first life was a rich pond of ingredients was formalized in the 1920s when a Russian, Aleksander Oparin, and a Brit, J. B. S. Haldane, both independently wrote about the emergence of complex biological molecules and life in conditions fuelled by an oxygen-depleted atmosphere on the early Earth. Haldane – a truly great scientist who would go on to become a central figure in the emergence of evolutionary biology in the twentieth century and a gifted science communicator – is an important character in this scientific journey as he first used the phrase 'prebiotic soup', and the idea of soup as the stock of life has bubbled away ever since.

Soup had its greatest moment in 1953, a vintage year for science. Crick and Watson revealed the structure of DNA in April, certainly the scientific achievement of the twentieth century. But at exactly the same time a young student was building on Haldane and Oparin's ideas to put together another, similarly iconic experiment. Stanley Miller was a twenty-two-year-old chemist at the University of Chicago, and as part of his Ph.D. begged his supervisor, Nobel laureate Harold Urey, to let him put together a rather fanciful experiment. He built a set of interconnected glass pipes on an electrified metal grid two metres square. This kit now sits in a dimly lit room in the labs of one of Miller's students, Jeffrey Bada, now an emeritus professor at Scripps in San Diego. It looks, not inappropriately, like a 1950s sci-fi experiment, with sparks and bubbling gases and coloured liquids. Miller filled the glass beakers with water, methane, hydrogen and ammonia, in an attempt to emulate what was believed at the time to be the essential ingredients of the early Earth. Following on from both Oparin and Haldane, Miller reasoned that the absence of oxygen in the atmosphere was key to the chemical maelstrom necessary to prompt the emergence of essential biological molecules. Miller put thousands of volts' worth of spark into that pipe-work, mocking the electrical storms and lightning that thundered from the turbulent Archaean sky.

Harold Urey had the good grace to let him pursue this experiment with the caveat that it would be shut down and he would move on to less implausible research if it bore no fruit within a few months. No such time was needed. Within days, the mix had turned first pink, then coffee brown. Miller extracted the rich brew, and his analysis of it found the presence of the amino acid glycine, and a handful of the other biological amino acids that are essential for building proteins. He published his results in the journal *Science*, noting in the methods that the conditions were designed to emulate the primitive Earth, not to optimize the production of amino acids. Just to consolidate that extraordinary result, there's a sweet coda to this experiment. In 2008, a year after Miller died, Jeffrey Bada rediscovered some of the original samples from this experiment tucked away in the back of a dusty drawer. He then subjected them to twenty-first-century analyses. Even in these fifty-year-old samples, precision inspection revealed not just the few that Miller had seen, but all twenty of the biological amino acids, and indeed five others too. It seems that under those conditions, the spontaneous production of essential biological ingredients was a trivially straightforward happenstance. Those simple units, it might be thought, would string together into proteins that all life depends on, and with their function the processes of life could begin. Miller had shown that in the tumult of the Archaean Earth the molecules that form the universal workers of living systems – proteins – were simply summoned into existence by the equivalent of a bolt of lightning.

The experiment was so appealing that Miller became an international celebrity. The press were gawping and excitable, and in their reporting rather extended the results to the point where some reported that Miller had created life. Of course, amino acids are not life, though they are essential for it, and their creation was big news. This experiment was seen as a stamp of approval for the idea of primordial soup, cementing its place as part of our culture, and the most tenacious idea in the origin of life. In a location somewhere on the Earth, a wet surface or pool or floating pumice was exposed to the gases of the Archaean atmosphere and a bolt of lightning. It carries a sense of drama, the spark of life injected into a corpse, invigorated from the skies, the moment of creation.

But could it have been like this? As we have seen in earlier chapters, the mechanics of life are mesmerizingly complicated. A cell is a hive of densely packed activity receiving input from its environment (whether this environment is as part of an organism or as a single free-living cell). Inside the cell, there is code that encrypts proteins, and those proteins enact the functions of living: feeding, communications with other cells and the reproduction of the organism to perpetuate the genes that it bears. To see the spontaneous emergence in a test tube of the molecules (or components of those molecules) that perform these vital acts certainly provides credence to the idea that the origin of life is neither mystical nor supernatural. At the beginning of the Archaean, simple chemical ingredients did make the transition from chemistry to biology, and Miller's iconic experiment follows in the direct scientific lineage of Darwin's warm little pond. In that charming speculation the recipe includes 'ammonia and phosphoric salts, – light, heat, electricity &c.', all of which are plausible components as they relate to some of a typical cell's processes. Miller's experiment tested this idea, basing the ingredients on a better understanding of the conditions of the infant Earth – ammonia, methane, hydrogen, water and lightning. In the sixty years since, many experiments have further refined the recipe, or shown similar and more sophisticated spontaneous construction of biological molecules out of a soup of ingredients. It's an attractive idea, and one that has stuck. But there are fundamental outstanding questions that underlie these experiments. Can life emerge by cooking up a chemical soup? Is a spark what it takes to drive a chemical reaction into a biological one? To answer these questions, and to get to the bottom of the origin of life, we must ask a very simple question, one with a deeply elusive answer.

4. What Is Life?

'I shall not today attempt further to define the kinds of
material I understand to be embraced within that shorthand
description, and perhaps I could never succeed in intelligibly
doing so. But I know it when I see it.'

Justice Potter Stewart, 1964

What is living? A definition of life might seem like the most fun-
damental bedrock on which a field as expansive as biology is built.
But, you might be alarmed to discover, there is no standard defin-
ition. At school, we get taught variations on a checklist for
identifying the characteristics of living things:
Movement
Respiration
Sensitivity
Growth
Reproduction
Excretion
Nutrition
Schools in the UK sometimes summarize this as the acronym:
MRS GREN. As a checklist, it works perfectly well for all the
living things we see around us: you're probably doing at least five
of those things right now.*

---

* For an even more reduced stock of animal behaviour, a more memorable but
less specific list is sometimes referred to as the 'Four Fs': Feeding, Fighting,
Fleeing and . . . Reproduction.

The criteria above are all facets of life, and are biochemical by nature – biology as enacted by chemistry. Movement, for example, can take many forms in living things. You scan these words when a network of specialized proteins contracts inside muscle fibres attached to your eyeballs, pulling and pushing the focus across the sentence. That's a very different type of movement to the daily bend of a sunflower as it tilts towards its solar power supply. That is controlled by a kind of inflatable joint at the base of the flower head, whose cells swell up with water on the side opposite the brightest light and bend the stalk towards the Sun as it traverses the sky. And that's different again from the movement of certain self-propelled bacteria, who come fitted with a beautifully evolved rotor, the flagellum, which spins at up to 1,000 rpm and can whip the cell along at a tenth of a millimetre per second.

We know very well that life is made of cells, and there are no life forms that are not built from cells. But that is not what defines it, in the same way that a house is not defined by bricks. We know that all life works via the reproduction of a universal code that we can translate and read, but you won't be able to understand a cricket match by reading the rulebook. That point where the Earth changed its status from 'vitally inanimate' to 'hosts living things' didn't happen by ticking off a checklist. There is certainly nothing wrong with that vital inventory. All of those examples on the checklist are by definition biochemical: the way proteins are built in strings and folded into three-dimensional thickets to give cells specific abilities; the way DNA can encode and replicate information; the way animal cells can take in oxygen, extract energy and pump out carbon dioxide; the way plant cells can take in carbon dioxide, extract energy, and pump out oxygen. And so on. All of these processes are chemistry in action, determined by the behaviour of the atoms that make up the molecules. So what is the nature of the chemistry that is actually biology?

Searching for a Definition

We have a checklist of the actions of life, but no clear singular definition. Alone, or even in multitudes, a protein is not alive, nor is DNA, nor is a metabolic pathway. These are, for desperate want of a less clumsy term, inanimate. They do not have the breath of life in them, though they are clearly essential for it. And yet we are in no doubt that the concatenation of all of these chemical events in an organism such as you is what allows you to live, and the ultimate termination of these chemical processes results in death. When we talk about the origin of life, everything pivots on the journey from inanimate matter to living matter.*

The transfer of information from cell to cell and from organism to organism via DNA is elemental to life as we know it. So is it possible to secure a definition of life that is underpinned by the ability to transfer information? NASA's interest in life extends from looking for it in space to synthetically engineering it here at home, as we can see in 'The Future of Life'. Astrobiology forms a satellite branch of origin-of-life research, the search for traces of life in far-off places. This new field is a coalescence of several branches of science, geochemistry, biochemistry, astrophysics, to contemplate the chances of life in the universe. NASA have three main priorities in this area: how did life begin and evolve; does life exist elsewhere in the universe; and what is the future of life on Earth and beyond? The first question – the origin of life – is of central importance to the astrobiologist as it will determine how we will look for and recognize life when we find it.† NASA's

---

* The word 'inanimate' is unsatisfactory as it implies inaction, which chemistry most certainly is not. The opposite of living – dead – is also unhelpful for the same reasons. 'Undead' seems a fitting way to describe what is more correctly called 'prebiotic' chemistry: fizzing, active reactions that form the pathway to life.
† They've been doing this for a while. In the 1970s, NASA sent the Viking missions to Mars, two insectoid landers that sampled the Martian surface looking

objectives in these off-world explorations has led them to adopt a definition of life in order to help specify mission parameters: simply put, 'if it's Darwinian, it's alive'. This is smilar to the position held by Jerry Joyce, a chemist working at the Scripps Research Institute in San Diego, whose experiments are precisely focused on allowing Darwinian evolution to occur in DNA's cousin – RNA. He told me:

> If there's a *sine qua non* of life it's the ability to undergo Darwinian evolution, and to have history in molecules. Chemistry doesn't have history. Biology has history. To me, the dawn of life is the dawn of biological history written in the genetic molecules that are carved through Darwinian processes.

This definition is focused in its entirety on information. DNA and RNA are codes, storing up a digital manual of instructions that can be copied and copied and copied. And as there is infidelity in that copying, they acquire errors, which in turn become new information, to be passed on, and selected if it is useful. That is evolution.* Replication is crucial for life, but is it enough to count as a definition? A crystal can grow, and can replicate its structure. That growth can suffer imperfections, but, according to this definition, it is not alive because those imperfections are not passed on. It is not Darwinian because it cannot acquire new characteristics via selection. Darwinian behaviour is certainly an essential feature of all life that we see, but is only one part of a set of behaviours that are universal in life.

Jack Szostak picked up a Nobel Prize in 2009 for a career working in human genetics. He helped build new techniques for us to

---

for the signatures of life, and found none. In August 2012, Curiosity, a rover the size of a VW Beetle, gracefully plummeted to the surface, with one item on its ongoing mission being to scratch around in the red rocks looking for traces of what once might have been life on Mars. It is the freakiest show.

* Joyce's research into this area is utterly crucial as he deals with the very origins of this process of descent with modification, and we'll come to it in the next chapter.

explore our own genome, and contributed to the discovery of key components of ageing. Then he switched from the human branch at one of the tips of the tree of life to the almost unrelated field of biology at the base of the tree of life. At Harvard, Szostak founded the Origin of Life Initiative. There, one of his many concerns is the spontaneous formation of the cell membrane, which, like life did, we will come to at the right time. He is quiet and easy-going, softly spoken, patient and thoughtful. But when I interview him, it's clear that the quest for a definition of life exasperates him, and more importantly gets in the way of researching it. Uncharacteristically, he *almost* snaps at me when I suggest that the absence of a comprehensive definition of life is problematic. He cuts me off mid-question: 'I don't think it's a problem at all. I think it's completely irrelevant. What we want to understand is the pathway, how we went from really simple chemicals to more complicated chemicals: from really simple cells to more complicated cells to modern biology. We just want to understand the process and all the steps. We don't need to say "Here's the dividing line: on this side there's chemistry and on this there's biology." The important thing is the path.'

J. B. S. Haldane, who earlier in the twentieth century had helped construct the idea of prebiotic soup, took a similar line in 1949, with his straightforwardly titled book *What Is Life?* Chapter 14 bears the same title as the book itself but begins with this bold caveat, brash enough for him to use capital letters:

I AM NOT GOING TO ANSWER THIS QUESTION. IN FACT, I DOUBT if it will ever be possible to give a full answer, because we know what it feels like to be alive, just as we know what redness or pain or effort are. So we cannot describe them in terms of anything else.

In other words, I know it when I see it.

Nevertheless, many people do focus on establishing a definition. Humans do like to categorize things, and science particularly so, because so often it aids comprehension. Edward Trifonov, a biologist from the University of Haifa in Israel, recently approached the problem with an unusual tactic, which was to look at the words used in the many attempts by scientists to define life. By throwing them into a melting pot, he then ladled out a purified phrase: 'Self-replication with variations'. This is similar to the NASA definition, and focuses on the transfer of information from one generation to another. I'm sure that this approach was well meaning, but it is hard to see the value in establishing a consensus definition based on the language used. The publication was gracious enough to include many responses from a cohort of scientists disagreeing and, in Szostak's case, resisting the urge to constrain life into a definition at all.

Indeed, all such attempts run the risk of 'blind man elephant syndrome'. The Buddha (but also Islam, Jainism, Hindu and other cultures) recounts the story of several sightless men asked by a king to tell him what an elephant is like. Each examined different parts of the beast with their hands and concluded that a pachyderm was only the bit that they could feel. The tusk examiner suggested it was like a ploughshare, the foot inspector a pillar, the tail man a brush, and so on. They fight; the king laughs. An elephant is all of those things, and you can't capture its majesty by singling out one of its characteristics.

When US Supreme Court Judge Justice Potter Stewart said 'I know it when I see it' (quoted at the top of this chapter), the 'it' he was referring to was pornography. The state of Ohio had banned the 1958 film *The Lovers* as an obscenity, but Stewart overruled. It has evolved into a phrase to describe the subjectively ill-defined, or things without clear parameters. Life is just that. At one point there was chemistry on Earth, and at a later point in time there were living things. The route from the first to the second point is inevitably long, tortuous and messy. The point where we definitely have living things is certainly where things have become Darwinian, but not solely (as we shall soon see, there are even certain molecules that display the Darwinian property of self-replication with selection). So the point is that the boundary between chemistry and biology is arbitrary. Life is a combination of lots of different chemical systems that are more than the sum of their parts. We separate science into categories in discussions such as this, and indeed when we learn it: biology, chemistry, geology, physics and so on. These are also somewhat arbitrary, as science is merely a way of knowing nature, and are particularly blurry distinctions when considering the very inception of one of them, biology.

The Energetic Fly in the Soup

All of the actions of cells are ultimately mediated by the controlled flow of electrically charged atoms from one side of a membrane to the other. As you read, charge-bearing atoms flow into millions of single brain cells until they hit a threshold. When this happens, the brain cell will fire up, and in consort with millions of others will form a thought process, or trigger a memory or understanding or induce the desire to make a cup of tea. Similarly, it is the controlled flow of protons – hydrogen atoms charged by being stripped of their single electron – across membranes that drives the generation of energy that the cell and organism entirely relies upon. In all complex life (including us), this happens in the cell's power stations, mitochondria; in bacteria and archaea across a membrane just inside the cell's outermost casing. This type of chemical pathway is part of what we call metabolism, and is of central importance to all life as it generates energy. That energy powers all of the actions of biology, including the ones that facilitate the replication of information across generations, and everything else on MRS GREN's list.

For this reason, in considering the foundations of chemistry as it relates to biology, we have to turn to a more fundamental science: physics. The way that chemicals behave is determined by the laws that belong traditionally to this field. Ernest Rutherford,* the discoverer of the particles that make up the atom, famously

---

* No relation.

declared: 'Physics is the only real science. The rest are just stamp collecting.' Though clearly a provocative gibe, there is some value in this reductionist view, typical of physicists. Biological behaviour is determined by chemical behaviour, which is determined by atomic forces, and these are in the realm of physics.

Haldane wrote *What Is Life?* in 1949, but he was not the first to come up with this deceptively simple title. In 1944, Erwin Schrödinger wrote a physics-focused biological treatise with the same name, a classic text drawn from a series of public lectures.* That this analysis should come from a physicist is perhaps fitting in further blurring the artificial boundaries between the modern disciplines in science. Physics, by its nature, tends towards the fundamental, and Schrödinger's conclusions derive from one of the absolute and most non-negotiable universal rules: the Second Law of Thermodynamics. This is the principle that dictates with total authority that over a period of time energy will always flow from a higher state to a lower state and never in the other direction. We see the application of the Second Law all around us. Once the heat is turned off, a pan of bubbling soup can only cool down. This principle extends to every aspect of our lives: the heat of a radiator is dissipated into warming our rooms because it is attempting to equilibrate the two imbalanced energy states: one is hotter than the other. It will never happen in reverse. The measure of the Second Law is what we refer to as 'entropy'. At a constant temperature, a plump balloon will only deflate, unless its knot is

---

* Schrödinger is best known for putting an imaginary cat in a box, and then poisoning it to death. Or maybe not. His famous thought experiment was a way of understanding the effect of observation on the quantum world and how it might affect the classical physical world. The cat, invisible to our observation, would be killed by the poison gas that has a random chance of being released. But we have no way of knowing whether or not the cat is dead until we open the box. According to quantum physics, until that point, the cat paradoxically occupies two simultaneous states: alive and dead, only to be selected upon observation. Perhaps less mindbending was his attempt to focus on the first of those states.

perfectly sealed, in which case it will remain constantly inflated. But it will never expand, if its surroundings remain unchanged. Within the closed seal of the balloon it has reached equilibrium and its entropy is constant. But relative to the rest of the world, it has a higher energy state and therefore wishes (if one can express desire in a balloon) to spread its energy more fairly. This tendency towards fair distribution in energy is represented by an increase in entropy.

What Schrödinger argued was that living systems are the continual maintenance of energy imbalance. In essence, life is the maintenance of disequilibrium, and energy as life uses it is derived from this inequity. It is sometimes described as a 'far from equilibrium' process. The entropy of the universe is bound to only ever increase, thereby ultimately creating a more balanced but less ordered existence. The temperature throughout the universe will eventually be the same, following the even distribution of its total energy as the Second Law of Thermodynamics commands.* Schrödinger recognized that all living organisms evade the decay to energy equilibrium during their lifespan, and continue to do so in their descendants. This has occurred continuously on Earth for almost four billion years. We consume food, and energy is extracted from it inside cells. In doing so, we construct an order within our bodies, without which decay would transpire.

This order maintained in living cells appears at first glance to be in direct contravention with the Second Law of Thermodynamics, which dictates that entropy will always increase, and therefore organization will decrease. Chaos is the ultimate direction of all things, and living things are not chaotic (at least on the terms laid out by physics). But this apparent paradox is not a problem. The law states that the incontrovertible increase in entropy occurs within a 'closed system'. The universe in total is a closed system, as

---

* At this point, universal entropy will be at its maximum, and the universe will have reached 'heat death'. But you can relax for now, as this will not happen for many trillions of years.

there is, by definition, nothing but the universe. On a more local scale, living things are not a closed system. We produce waste as a result of our life-sustaining metabolism, and expel it into the rest of the universe. Whilst order is increased and maintained during any lifespan within the living thing itself, this seeming contradiction of the Second Law is more than compensated for by an overall increase in entropy beyond the confines of that organism: that is, your waste. The entropy of the amount of waste you have generated in your life is overwhelmingly more than the reduced entropy that your body maintains by being ordered. And thus, the laws of the universe remain perfectly intact.

Life is a process that stops your molecules from simply decaying into more stable forms. The process of living is the chemistry that perpetually holds off decay. And this is precisely why the concept of the primordial soup is flawed. The notion that the right ingredients in the right surroundings might generate a self-sustaining life form ignores the underlying principle that life is a far-from-equilibrium process. The chemical activity in a soup can only accede to the Second Law of Thermodynamics: unless it has an external source to maintain an energetic imbalance, it will only decompose. In Stanley Miller's experiment, the spark of lightning may have triggered the formation of amino acids, but did not power a system of disequilibrium. Once those chemicals had reacted, they would do so no more.

Bill Martin is one of the main critics of soup-based origin of life science, and we will come to his work shortly. He proposes an easy experiment to challenge the concept of primordial soup: mash up a life form of your choosing to the point where any cellular resemblance is destroyed, but all of the ingredients are still intact. This experiment effectively occurs every time a cell dies, but spontaneous resurrection from this soup, with all of the right ingredients present, remains a myth. Any model of the very origin of life that does not take into account the necessity of a continuous flow and manipulation of energy is building upon something that is already dead. Stanley Miller's iconic experiment remains important, though its comment on the origin of life is limited.* It shows, elegantly and incontrovertibly, how biomolecules will arise from basic chemistry in the right conditions. However, it reinforces the view of life as merely an assembly of chemicals contrived into a thing that can reproduce. A primordial soup is not a vital and energetic mix as it has no way of sustaining an imbalance of energy, whether it is in a warm pond, a pumice raft, a muddy volcano or any of the other locations that have been proposed for the origin of life. A primordial soup is condemned as a midden: a decomposing dump.

In a sense, we living things are out of step with the rest of the universe. In the *Origin of Species*, Darwin described the 'struggle for existence', meaning the fight to get food, or a mate, or endurance against the elements. But it applies at a more basic level. To be alive is to struggle against entropy. Life does not violate the Second Law of Thermodynamics, not at all. We cannot beat it, as that is the invincible force of scientific laws. In death, we all submit to the will of physics, and our atoms accept their universal fate: to decompose, and to be recycled, ultimately into less energetic states. Entropy strives to make the universe both more chaotic and more balanced. In doing so, entropy always increases; just like in casino gambling, the House always wins.

But by being alive, we have the opportunity to take something back from the House, or at least slow down its inevitable victory, at least for a while. Life has evolved to extract energy from our surroundings and use it to maintain our vital information against the universal slide towards equilibrium by swapping and pumping protons from one side of a membrane to another inside a cell's innards. Our lives, all lives, conspire to manipulate the fundamental forces of nature, and strive to do so continuously and forever. Jack Szostak is probably right, and tearing our hair out trying to encapsulate the essence of a life is merely a distraction from trying

* The gases of Miller's experiment were probably not the ones present in the early Earth: it's now thought that carbon dioxide flooded the skies, and nitrogen was also present. That's a side issue here; we will never know what the exact composition of the early Earth was.

to trace back the path we know it took. But this indulgence into physics is crucial for understanding the origin of life. We know we can't go back in time and observe it. How it happened once is lost to science and history, so we do what we can to emulate it in conditions that gave rise to life's enduring imbalance. At some point, that imbalance acquired or created a system that allowed energy capture to be sustained independently. In that hatchery, the beginnings of encoded Darwinian descent could begin. But life forms are first and foremost a sophisticated collection of chemical behaviours underwritten by the need for energy, and that informs how we go about experimentally quizzing the origin of life. In other words, we need to hunt for the signatures of metabolism.

It is clear that Luca was a living thing from which all subsequent life arose, but already carried with it many features that are essential components of the life that followed, including metabolism and genetics. In our cells, there are two parts to metabolism: the first is the digestion of molecules to release energy, and the second is using that fuel to make life-sustaining molecules, including DNA and proteins. That poses something of a conundrum. From Luca onwards cells do the things on MRS GREN's list, and we will come to the assembly of such a sophisticated system in the final chapter. But each one of them is dependent on a chicken and egg problem. DNA encodes the proteins that enact cellular functions including metabolism, and these functions enact the decryption of the code itself. We've seen how the code works and how it was discovered. We've seen how it is the backbone of evolution, and the template for the richness of all life. So how did DNA come to be?

5. The Origin of the Code

'For all inferences from experience suppose, as their foundation,
that the future will resemble the past.'

David Hume, *An Enquiry Concerning
Human Understanding* (1748)

Hebrew has twenty-two letters. English has twenty-six, and Sanskrit fifty-six. Chinese languages use pictograms, and these number in the thousands, depending on how you count them.

Life only has four letters in its alphabet: A, T, C and G. Throw in a few extra glyphs and the flexibility of this alphabet goes up a touch. Linguists call these tweaks 'diacritics': the circumflex in French (â) or the German umlaut (ö). DNA's diacritics come in the shape of a small molecular appendage, methyl, which is simply three hydrogen atoms bound to a carbon atom, and gets tagged onto A or C. Just as in languages, these annotations are an essential feature of genetics, modifying the meaning by labelling sections of genome for particular characteristics. Most significantly, this earmarking of the code marks out areas of DNA to be silenced, as if leaving the text where it lies, but scoring through entire blocks to say: 'Do not read this sentence.'

Even with this type of flagging, the most generous number of letters in living code is way lower than for even Hebrew. Evolution has given us a comprehensive description of how the wild spectrum of species blossomed from this simple code, but very little about how it came to be. When pondering the origin of life, however one tries to define it, at some early point the origin of our alphabet is a key question.

It's not just the alphabet that is bafflingly conservative. It combines into a similarly limited vocabulary. As explained in Chapter 2, in genes, those four bases are arranged into triplets, each one spelling out an amino acid. When strung together, these form proteins. But there are only twenty amino acids encoded in the genes of all life forms. Furthermore, the way amino acids are encrypted in DNA is studded with redundancy. There are sixty-four ways of arranging four letters into groups of three. Sixty-one combinations are used in genetics to spell out just the twenty amino acids (and three indicating a 'full stop' signal to mark the end of a protein). This means that many triplets encode the same amino acid. For example, TTA spells out an amino acid called leucine; but so also do TTG, CTC, and three other variations.

This redundancy allows changes to be acquired in our genes that don't make potentially harmful changes in the proteins that they encode. If during the process of a cell splitting in two a random DNA error switched TTA to TTG, it would still spell out leucine, and the protein containing that amino acid would be unchanged by this unforced error. If it changed the final A to a T then leucine would be replaced with phenylalanine, an amino acid with similar properties. So it would potentially change the nature of the protein, but perhaps not too dramatically.

We see some similar redundancy in language. Here in England, your favourite colour might be blue. But 6,000 miles west, blue might alternatively be your favorite color. Through random mutation, almost certainly a copying error at some forgotten point in history, American English has excised a letter that we Brits believe is essential. It is patently not, as the pronunciation is the same, and it's not difficult to imagine a world where my British extra u slowly degrades out of use altogether.

That's not to say that all mutations are equally benign. Drop the r from 'friend' and you have a companion with a somewhat opposite character. It may not feel like it if one is unfortunate to suffer from a genetic disease, a category that includes all forms of cancer, but, given the amount of DNA replication that goes on, harmful mutations are relatively rare. But when they do happen, a single letter change can have catastrophic consequences. A single base change in DNA that results in a switch from one amino acid to a very different one may well cause problems. Switch the A to a T at a specific point in the β–globin gene and the amino acid switches from glutamic acid to valine, which has quite different chemical properties. The result is a misshapen globin protein, which in turn deforms the shape of red blood cells into bent and elongated curves instead of the concave round lozenges that they are meant to be. The host of this single letter copying error will have the disease sickle cell anaemia, a blood disease that frequently causes premature death. Such is the harsh but mercifully unlikely nature of genetic disease.

For the most part, though, these errors are of little consequence. They are normal and happen every time a cell divides. In copying of the whole genome during replication, proofreading does occur. The copying proteins, called DNA polymerases, check that the new strand they are making matches up with the template, ensuring that wherever there is an A, a T is set down, and not anything else. But it's not perfect, and sometimes during cell division new changes make it through the proofreading. If that occurs in a sperm or an egg, it can be the first step in the genetic Chinese whispers that drive evolutionary change. You are different from your parents not just because whole genes get shuffled around during sperm and egg production, but because those cells bear new single changes in their DNA, which will be random, and therefore unique to you alone. Sometimes, the pairing of the As and the Ts, the Cs and the Gs goes awry, and one of the two strands of DNA bulges out like a mismatched zipper. If uncorrected by the proofreading proteins, this single change can be passed on, and may slightly alter the behaviour of the protein it codes.*

---

* The frequently quoted stat that genetically all humans are 99.9 per cent similar is of relevance here. That means that, if we compare two individuals' DNA, they differ by only one letter in a 1,000. But if you factor in that there are three billion letters of code in a human's genome, that makes three million individual

How is it that we have settled on four letters, arranged into triplets? With such a neat self-contained system, unpicking it is tricky, and knowing where to break into the circle is perplexing. One approach, though, is to work out the minimum amount of DNA required to encode the twenty amino acids that life uses to make all its proteins. If there were only three letters of genetic code, then there would be twenty-seven possible combinations of triplets, still more than enough to encode the twenty amino acids we need. But in reducing the redundancy by more than half, the buffer against potentially harmful diseases is also reduced. The three letters in the triplets are not all equal. We see patterns in the ordering of the bases in triplets and the amino acids they encode. The first base relates to the process of the source of the amino acid. Amino acids float freely in the cell, waiting to be assembled into their codified proteins.* Some are the products of the cell's metabolic cycles, but nine of them we cannot make, and have to eat. By comparing triplets with the same first letters, we can deduce their origin as being self-made or eaten. The second letter corresponds to a type, options including water-loving (they dissolve easily) and water-hating (they don't). The first two letters have a clear purpose in determining the product. The third, with all its wildcard flexibility, secures the deal, locking it into just one of the twenty. We can therefore speculate that the first form of DNA code might have been not a triplet but a doublet, fixing the deciphering to essential sets of manufacture processes. The addition of a third allows both more combinations and more variation in the sequence,

---

letters different (and doesn't include all manner of different, larger variations within our codes). That is a lot of variables to play with, and goes a long way to explaining why we are all unique, even identical twins. Your precise genome sequence has never existed before and will never exist in another person. Humankind is in the 99.9 per cent, but you are encrypted in the wealth of the remainder.

* A more detailed description of this key process is described on p. 100 in 'The Future of Life', as scientists are beginning to subvert and redesign this process to create new, unnatural proteins.

creating a conservative code that not only protects against the impact of catastrophic change, but gently encourages subtle modification. In short, our DNA encourages evolution.

But the underlying code is frozen, without change for probably four billion years. Alphabets are ice-cold, but not frozen. They do change, but the change is imperceptibly slow. DNA, though, is indeed locked, at least in nature, though that is changing in the new world of invented biology, as you can see in 'The Future of Life'. Francis Crick once thought that the alphabet of DNA came to be fixed as a result of a 'frozen accident': a system that worked well enough, either it out-competed other versions or they didn't ever exist. But it is now clear that the inequity of the letters in each triplet is no accident, and became frozen with a delicate balance between fidelity and mutation, like a loving parent encouraging children to explore but at the same time protecting them from harm.

Ghost World

We can begin to see how DNA might have originated in a simplified form and settled into its current stable existence. But the paradox here is much more complicated than the old problem of the chicken and the egg.* Copying DNA is dependent on proteins, and proteins are encoded in DNA. DNA is the code and protein is the active product. But there are very solid reasons to think that the beginning of life, the origin of genetics, began not with DNA, but with its simpler cousin RNA.

Again, by looking at how modern life forms work, we can infer earlier, cryptic roles for the mechanics in cells. RNA is the middle part of Francis Crick's misnamed 'central dogma': DNA makes RNA make protein. In order to understand how this sequence might have come about, it's useful to think of it in steps. We know

---

* Obviously, egg-laying reptiles came hundreds of millions of years before the evolution of any birds, let alone *Galus domesticus*, the humble chicken.

of no way to directly make protein from DNA without the go-
between RNA, so we might suppose that the first part, 'DNA
makes . . .' might have been added after the concluding step 'RNA
makes protein', RNA being the coded transcript from which the
proteins are made. RNA, a single strand, is less chemically stable
than DNA with its paired helices; it is more prone to falling apart.
DNA retains the code for proteins with one strand reflecting the
other, providing a mirrored backup service: where there is an A
the other strand holds a T, and where one holds a G, the other
bears a C. So it's not unreasonable to imagine that DNA emerged
later as a more secure form of data storage.

More recently, a pathway for that transition has even been suggested by looking at how error-prone DNA and RNA are when
copying one to another and back again. A Harvard team lead by
Irene Chen compared how faithful the copying is when making
RNA from RNA, RNA from DNA, and DNA from RNA.
This is akin to testing the prowess of an Internet translation engine
by inputting a sentence, translating it, and then back again to
see how mangled it gets. Unequivocally, using DNA as a template
came out on top. When copying RNA from a DNA template, the
transcript was most faithful. This suggests that a move from an
RNA world to the one we have today could have occurred without static – the message would be preserved. But DNA copied
from RNA was riddled with errors.* With this concept in mind,
it seems reasonable that as an information-storage device DNA is
more robust, more secure and more faithful than RNA. The transition from a world where RNA was the information bearer into
the biological era that we know is referred to as 'genetic takeover',

---

* For example, the sentence 'But DNA copied from RNA was riddled with
errors' is translated into Icelandic by Google Translate as 'En DNA afrita frá
RNA var riddled með villa', and back into English as 'But DNA copy of RNA
was riddled with errors', which is inelegant but pretty close. Repeating this
experiment with Czech results in 'Ale DNA zkopírován z RNA byla
prošpikovaná chybami' and then 'But RNA copied from DNA was riddled
with errors', which is exactly the opposite of the original meaning.

and it seems that, once that annexing had occurred, we could never go back.

That chicken-and-egg conundrum – DNA codes, protein acts – is at least partially solved with RNA as well, as on occasion RNA can do both. A second clue for thinking that there was a world sporting the signs of life that was populated by RNA, not DNA, comes from the cell's protein factory, the ribosome. When it is time for a gene to make a protein, the process goes like this: the gene, set in DNA in the host's genome, receives an instruction to be turned on. A protein unscrews the double helix, like unwinding a wire cable tie, and separates the two strands. Another protein clamps onto the coding strand (the other is a mirror, and performs minimal function), at the three letters ATG. This codon is for the amino acid methionine, but also marks the beginning of a gene, the 'start codon'. From there it walks mechanically forward, copying the DNA into a mirrored RNA molecule: where it sees ATG, the RNA will read UAC.* When the transcription of the DNA is complete, the wispy RNA molecule floats off, bearing the message of the gene, and is helpfully called 'messenger RNA'. The ribosome then picks up this transcript and ingests it, one letter at a time. It reads the code three letters at a time, each codon specifying an amino acid, which is delivered to the ribosome from the cell's milieu. As each codon is read in turn, the ribosome picks up the amino acids and strings them together to make a protein, which is expelled into the cell, dispatched to serve its purpose.

The ribosome itself is made of several parts, just as many of the active parts of life forms are.† These smaller parts self-assemble far more easily than flat-pack furniture, by jiggling themselves into position. But what is interesting for our purposes is that more than

---

* RNA uses uracil as a replacement for thymidine, which differs only by the presence of a small cluster of atoms called a methyl group, also used as the diacritic tag in DNA methylation.

† For example, the transporter of oxygen around our bodies, and indeed what makes blood red, is called haemaglobin, and is assembled from four units of protein parts called globin, around an atom of iron.

half of those parts are made not of protein, but of RNA. These long, heavily folded slivers of RNA link together with proteins to make the working ribosomes, and function just like proteins in that they enact a process. So with these types of RNA molecules, we have both information and function. With RNA as the fore-father of DNA on the early Earth, the paradox of DNA and protein – the former encoding the latter and the latter making the former – vanishes. This idea is referred to as the 'RNA World hypothesis'. We can get round the chicken-and-egg problem by not needing either. At some ancient point on the road from mere chemistry to biology the central dogma of 'DNA makes RNA makes proteins' was simply 'RNA makes'.

The First Photocopiers

NNNNNNUGCUCGAUUGGUAACAGUUUGAA
UGGGUUGAAGUAU–GAGACCGNNNNNN

Can you see the family resemblance? This is R3C, and is the child of Jerry Joyce and Tracey Lincoln from Scripps in California. If evolutionary genetics is the process of tracking the slight changes in our genes back through time to reconstruct the Darwinian rep-licators of the past, R3C might be the end of the line.

With RNA being a decent contender for the first world of information and replication, this is a simple piece of RNA that does both those things. The letters are standard RNA (*N* being a wildcard, representing any of the four bases A, C, G and U), and is made of two parts (shown separated by a dash). In a test tube, R3C contorts into a shape like a hairpin. Its function is to makes a mir-ror version of itself by linking the two parts together. This propinquity for its reflection in turn makes a new copy of the ori-ginal, and so on. This goes on ad infinitum, as long as the system is fed with the ingredients that will enable the continued chemical reactions that drive replication. Hundreds of millions of copying molecules can be made in a few hours.

It is a kind of proto-gene, made of RNA not DNA. Unlike our own genes, which have functions and instructions for proteins that build tissue and bone, or issue commands so that other genes will do so, the instruction that R3C carries is simply 'copy myself'. DNA relies on other mechanics to copy itself, but R3C doesn't need any help: it's a photocopier that only makes other photocopiers. The fact that its only instruction is 'copy myself' doesn't make it very useful for a modern organism. But genetics had to start somewhere, and, hypothetically, this might be similar to what the first genes looked like – a copying machine. RNA that has function is called a 'ribozyme'.* It's in these simple double-action molecules that many scientists see a fundamental precept of life: reproduction of information. Joyce believes that this is 'where chemistry starts turning into biology. It's the first case, other than in biology, of molecular information having been immortalized.'

In biological terms, this is the smallest unit of information, in computer terms, a single bit. But as they promote their own replication, in the test tube, they enact a form of Darwinian selection. Joyce's experiment at first was marked with fidelity: the ribozymes reproduced themselves impeccably. But evolution requires flawed copying, and, as Joyce tells me, 'perfection is boring'. So they introduced imperfection into the mix, sequence infidelity – that is, spelling mistakes – in order that each copy was potentially fractionally different. Hence Joyce introduced the foundation of Darwinian selection: variation. Out of a pool of these short RNA molecules, ones that reproduce themselves with greater success emerge to become the dominant form. These molecules have, through no

---

* A portmanteau of ribo- for RNA, and -*zyme* from enzyme, proteins that catalyse biological reactions. It is annoyingly similar to the word *ribosome*, which itself has ribozyme-like properties, that is, RNA with function.

guiding hand other than Joyce's providing the right conditions and feedstock, undergone an unliving form of natural selection.

It's a clever approach to the origin of the genetic code. These experimental ribozymes are not natural but they are only partially designed. David Bartel and Jack Szostak created a technique in the 1990s that gave a great fillip to the idea of an ancient RNA world before DNA and proteins. Their technique allows functional ribozymes to at least partly create themselves. It's akin to the fabled monkeys with typewriters idea, that with a big enough number of simians bashing away on keyboards, one will eventually type a Shakespearean sonnet. Bartel and Szostak put together a pool of literally trillions of strings of RNA which were identical for a short stretch at one end, and then random for the next 200 letters. In the trillions of possible combinations they were looking to find one that, purely by chance, had the ability to add another similar RNA molecule onto itself. They then fished in that pool with a tagged piece of RNA, a bait to hook any one of those trillions of random molecules that could clasp another RNA molecule. In the monkey typing metaphor, it would be like searching trillions of simian text files with the term 'darling buds of May'. If that sounds easy, remember that these are entirely random sequences, and they are asking for meaning, or in this case function, to come from nothing. Bartel and Szostak found exactly that ability from their pool of RNA molecules at a rate of around one in twenty trillion.

The monkeys and typewriters example is purely random, and makes the point that, when working with very large numbers, patterns (or prose) will emerge.* Evolution is very much not random. The variation in genetic code may have arisen entirely by chance, but selection (whether natural or by the hand of a creator) is quite the opposite. Bartel and Szostak fully emulate biological evolution with their method, by allowing variation to occur, but

---

*In 2003 researchers at the University of Plymouth tested this idea in a very small-scale experiment. Six macaques were left with typewriters for a month, and produced five pages consisting mostly of the letter S. However, they also destroyed the typewriters by urinating and pushing faeces into the keyboard.

selecting the variants that work. Then, just as in nature, they repeated, but this time using only the ribozymes that already had demonstrated an RNA linking ability (as if, hoping not to stretch my analogy too far, giving the monkeys the first line of a sonnet as a starting point to their frenetic random typing). After ten rounds of fishing for those that could bind RNA together, the ribozymes that had made it through this harshest of talent contests were several million times better at RNA joining. There's a drop of tonic to take with that, which is that naturally occurring enzymes work at a rate that is around 10,000 times faster than the best of Bartel and Szostak's ribozymes, which were unlikely to improve much more. But recall that we are dealing with introducing function into a world where there was none, and this RNA world was transient, destined to be replaced by the more efficient and effective DNA and proteins.

In essence, these experiments are attempting to create something more fundamental than life. That ribozymes were the first genes, needing neither protein nor DNA, is not something we can actually know for sure, though it is uncontroversial now. Ribozymes do exist in nature, but if RNA self-replicators were the first genes, they've been extinct for more than three billion years, and we are experimenting to resuscitate an unrecorded dead language that is only known by its descendants.

In Cambridge, UK, the Laboratory of Molecular Biology has been called the 'Nobel Factory', as nineteen of those most sought-after gongs were won for work done in a dull grey glass building, now relocated round the corner to a shiny new grey glass building. There, Philipp Holliger runs a lab where they are exploring the lost RNA world. His crew used a similar test tube evolution to pick out another candidate for an aboriginal gene. They started with a smaller lucky dip, a paltry ten million variants, and, by using some technical wizardry involving oil and magnetic beads, selected a ribozyme that produces more RNA. This one doesn't replicate itself, but they did manage to cajole it into producing an entirely different ribozyme, a well-known one rather more dramatically called the 'hammerhead'.* So far, Holliger's ribozyme can only make new RNAs of around 100 bases in length, which is a little short for a ribozyme, and a long way from the average modern gene, but they are working on extension.

Here we have the emergence of plausible models for the origin of information in chemistry and, more importantly, the origin of information copying. This is a hallmark of life. The amazing ability of Joyce and Holliger's ribozymes is their spontaneous function. Joyce's copies itself, and Holliger's copies lots else. The great goal is to shape the emergence of one that does both.

Creation by Deletion

But we can go back further. Ribozymes, as with all genetic code, have their four letters with which they harbour information, A, U (instead of T in DNA), C and G. If we assume that Luca had this alphabet in place, and that is as far back as our historical records can go, we have no record of how this complete code came to be. This is unlike the evolution of language, where we have historical records of previous forms, with different letters and meanings. The acquisition of the language of DNA and RNA is crucial to the understanding of the origin of life. To think that all four letters of code were acquired simultaneously seems more unlikely than if they were acquired sequentially, one at a time. We can test that idea by deleting them sequentially.

As in Scrabble, some letters are more valuable than others. C, cytosine, bound to a G is more stable than a bit of code containing A and T, which will be the first to fall apart in heat or cold. On top of that, in RNA, the letters can pair up in a way that is slightly different from the neat and tidy rungs of the DNA ladder. This,

---

* The hammerhead ribozyme is found in a number of viruses and single-celled organisms. Its name is derived from the fact that, when the RNA sequence folds into a 3D structure, it has a head that is almost exactly the same shape as a hammer.

rather sweetly, is called 'wobble pairing', and helps the RNA fold into the loops and folds that give ribozymes their self-copying ability. In the absence of C, both normal and wobble pairing can still occur.

Again starting from a colossal pool of ribozymes with random variations in their sequence, Jerry Joyce and his colleague Jeff Rogers drove evolution by repeatedly fishing for ones that contained no cytidine,* using a bait with little affinity for that letter, but that could still join up two RNA molecules. Effectively they were breeding out a particular trait from their stock, but instead of its being an undesirable characteristic in an animal or plant, the stock is a 140-letter ribozyme, and the trait is the letter C. Even in the absence of one-quarter of the available letters, they bred a ribozyme that could ably retain its joining function.

So, once you've discarded one of the letters and showed that biological molecules can still work, what is the next experiment? Joyce's approach, obvious but no less brilliant, was to evolve a ribozyme with only two letters of code. Necessarily, this involved evolving ribozymes with similar but different letters, in this case referred to as D and U, and the activity of the molecule they bred was much reduced from a full alphabet version. But nonetheless it still functions as an active biological tool. It may even be the case that this limited alphabet molecule would have been advantageous in hot Archaean waters, as the folded bonds that contain C might have self-destructed at high temperatures.

These clever experiments don't show what actually happened in the origin of genetics, but they do show what might have happened. They show that evolution can proceed with an alphabet that is more restricted than the one life now uses. They feed into a framework that persuasively builds up an origin for coded replication, that is, genetics. If one of the defining characteristics of life forms is the ability to store and reproduce information, then one

---

* Cytidine is the name for cytosine attached to the ribose sugar, that is, the rung attached to the strut.

of the most fundamental questions of the origin of life is how such a complex system could have begun. How it compares to what might have happened four billion years ago is difficult to say. We have the advantage of ingredients and design, at least experimental design, rather than messy bucket chemistry of the early Earth. But that first time had millions of years, whereas scientists have made these new replicators in a decade. The task is not to replicate what happened once. That will be impossible. But in all origin of life studies it's important to remember that we know the answer: life is the answer. The question is finding a believable route to get there, and in these experiments that is beginning to emerge.

The Origin of the Alphabet

And we can go back further still. When Jerry Joyce asks one of his brews to copy and mutate ribozymes, he has to supply the ingredients. The Archaean Earth didn't have a clean lab with sterile glassware and manufactured chemicals bought purified from industrial companies. His ribozymes provide a plausible mechanism for the first genes, and the basis for a language that has survived for billions of years. The letters of this language are the bases of RNA, so the next question is: where did they come from? These are not trivial molecules, certainly in terms of their importance in bearing the code of life, but also because they are complex. At least we describe them as complex molecules for a couple of reasons. One is that the letters of genetic code are manufactured in a difficult-to-learn biological pathway involving many different proteins. Our cells do this unthinkingly, of course, as they have evolved highly advanced processes of metabolism that sustain their own existence by making their own parts. We look on these processes and carefully work them out with awe, and sometimes terror, when revising for exams. The metabolic pathways that create bases in our cells are evolved over innumerable iterations – the most stringent, patient and ruthless design process known. But of course, if these complex chemicals predate the seed of life's tree, there was no sophisticated metabolism with which they could be built.

With that in mind, the chemical structure of the letters of genetic code are complex primarily because we say so. If that sounds like circular reasoning, the second reason we think of them as complex is because they're not easy for us to synthesize chemically either. The cell's process of synthesis is intricate so we see it as complicated, and our attempts to replicate that synthesis are tricky as well. Those bases have to line up in exactly the right way to form a working ribozyme, and for decades this has proved to be largely elusive in the chemistry lab. Additionally, there is a major supply problem. To make one of Jerry Joyce's self-replicating ribozymes you need seventy-odd RNA bases, but that's if you know exactly what you are making. If one were to emerge from a random pool of millions, you'd need billions of bases to begin with. And when it starts copying itself, the number of letters goes up exponentially. Ten rounds of replication of a 100-base ribozyme would need a pool of more than 10,000 bases, but 100 rounds of copying would need more than 10^{30}. Add to this the fact that each round reduces the concentration of the letter pool, which makes the replication harder. It would be like spelling words out of Alphabetti Spaghetti. After using up a few letters, it becomes harder to spell more words. So in order to keep spelling out new words you need a constant supply of letters, possibly an entire Heinz factory.

So before we get to the self-catalysing, self-copying ribozymes, an earlier, more basic problem is where those ingredients come from in the natural, inanimate, embryonic world. The problem is this: without biochemistry to hand, how did these letters spontaneously form, when we find it so difficult in the lab? This has been a question for the entire duration of the RNA World Hypothesis' life, stretching back nearly forty years.

If DNA is a twisted ladder, the two essential components are the rungs and the struts. In RNA, which is only a single strand, it is like a ladder cut vertically in half. In order to get the individual letters of code to link up into a working molecule, they have to link up to the struts of the ladder. This third component is a linker made of phosphate – a combination of phosphorus and oxygen atoms. The struts, meanwhile, are made of types of sugar molecule: the R in RNA stands for the sugar ribose, and the D in DNA for deoxyribose, which is the same except for one less oxygen atom. Each molecule of these sugars is attached to one of the letters, A, C, G or T (or U in RNA), and when these sugar molecules are stacked up, they form the backbone, the ladder's struts. In DNA two backbones pair up, in RNA they remain as a single strand. It is the shape of these molecules that determines their behaviour. For example, uracil is a small hexagonal ring of atoms, and it links up with its sugar, ribose, which is a pentagonal molecule like two panels on a football. Together, they make up a single rung of RNA, and these will get linked together by phosphates with the other bases to make a piece of genetic code. This brief aside into possibly painful chemical structures helps to flag up why re-creating life anew is a tricky business. Only these precise shapes will act as part of living cells, but there are plenty of other ways these molecules could be rearranged. An atom in the wrong position, a molecular appendage in the different place and you don't get DNA or RNA, and you don't have life.

The trouble is that biological chemistry in cells just happens, whereas we have to try quite hard to emulate it. While we can observe and deconstruct the mechanism that cells use to manufacture these molecules, for the purposes of studying the origin of life we need to understand their synthesis before those manufactories were present. Outside the foundries of our cells, getting the rings of sugar and uracil to link up has proved to be a pain.

Chemistry is an ancient science, and has well-worn pathways of synthesis. Making complex molecules is frequently done by laying one stone at a time, building up to the final product. Some steps are harder than others, and atoms have to be cajoled into forming bonds where we want them to be rather than somewhere easier.

Gentle coaxing can persuade two reluctant atoms to link up on the pathway. But in the case of making uracil for use in RNA, this stepwise chemical synthesis has proved unfruitful, as it has a reluctance to link its two component rings.

John Sutherland and his team are also at the Laboratory of Molecular Biology in Cambridge. His approach to this problem is a bold step, at least in terms of chemistry dogma. In a landmark study in 2009, they successfully bypassed the obstruction by taking an entirely different route.★ The traditional approach is to keep the production of the two components – the ribose rings and the uracil ring – separate, as both processes produce a tar of products. Mixing them together, it was assumed, would result in a chemical sludge containing many sugars, and a lot of other stuff, but not much uracil. To get round this, Sutherland ignored the prejudice of that stepwise pathway to sludge, and decided to mix the ingredients together in the first place. This technique is called 'systems chemistry', where instead of assembling the parts of a complex molecule one after another, and purifying after each step, all of the ingredients are mixed at the same time. It is in some ways reminiscent of Stanley Miller's experiments in the 1950s, which produced amino acids in abundance from a stock of ingredients supposedly present in the young Earth. It's categorically different, though, in that it is not merely mixing together ingredients to see what will emerge. The production of uracil in this way is heavily designed to do exactly that, in an environment that conceivably was present. It's not often that scientists use a word such as 'plausible' in the title of a paper, but in Sutherland's case it fits their model: 'prebiotically plausible conditions'. I have been dismissive of the idea of primordial soup in the previous chapter, on the grounds that it is not energetically plausible that a self-sustaining living system could emerge from a chemical swill. This is different, as it is a

★ This work was conducted while Sutherland was at the University of Manchester, a Nobel factory itself, boasting more than twenty laureates.

mechanism for building up the elements of living things, in this case the language of genetics.

Sutherland's approach works – uracil is formed – and is summed up with his comment 'Complexity is in the eye of the beholder.' His reasoning is that it must have happened in the past without the assistance of a chemistry lab or the cellular machinations of biology. This new pathway is itself of great importance, but somewhat reserved for chemistry wonks. It turns out that the key missing seasoning was phosphate, which had the effect of jamming sugar production, but nurturing the linking of the two rings to form uracil.

But the approach says something significant about attitude in science. Sutherland's thinking is rather radical. For the first time, instead of thinking about the chemical conditions of the early Earth and trying to emulate them to make uracil, his team figured a way to synthesize uracil and then postulated that these conditions were likely to have been the chemical swill from which code built itself. After all, we already know the outcome on the early Earth: it happened, regardless of how hard we find repeating it. Uracil and all the elements of genetic code were created somehow and somewhere, and if we can re-create that, we potentially know more than we did about what the early Earth was actually like.*

Our prejudices about the difficulty in the non-biological creation of complex molecules was also challenged in 2008 when a team led by Zita Martins at Imperial College London isolated uracil in the Murchison meteorite. That 100 kg rock had fallen to the Australian earth in 1969, and has been subject to scrutiny ever since, not least because it's big, which means there's lots of material to work on. It's also full of carbon, and so of interest to

---

* There's a coda to Sutherland's synthesis. They found that shining ultraviolet light onto the mix had the double effect of improving the yield of uracil and of carving up some of the by-products. UV light on Earth comes from the Sun, stronger now than it was when the solar system was young. This, of course, is a plausible addition to the Archaean reaction, as these conditions were present on the young Earth, but makes the idea more focused, at least in a location. If this indeed transpired, then it happened on the Earth's surface.

astrobiologists, many of whom suggest that extraterrestrial rocks might have delivered some if not many of the chemical components to Earth. Certainly, other bases and amino acids have been found in meteorites, and, given the flow of these boulders from space during the Late Heavy Bombardment, this might be a plausible supply. Either way, the presence of uracil off-world shows again that its synthesis is perfectly possible, no matter how hard we find it to do.

It's difficult to shake people's prejudices, but there is no place for dogma in science. When explaining his new pathway, Sutherland faced vocal criticism from entrenched chemists decrying the simultaneous formation of the ingredients, insisting on the traditional view that synthesis should be incremental and stepwise, not derived from a chemical brew. He now jokes that this criticism is the 'Dem Dry Bones' model, in which each element is created ready to be assembled into a whole: ' "The foot bone connected to the leg bone," so how does evolution produce a lower limb? By connecting a preformed foot and leg of course . . . "Now, hear the word of the Lord!" '

With these experiments, the emergence of the universal language of life is not solved, but they reveal a pathway that looks rather plausible. The language of life is a staggering example of natural design. Not only does it have the ability to encode the wonders of the living world, it is built with a natural buffer in it, one that encourages evolution, but not too radically. We now have a grasp of how this code came to be. We have credible roots from basic chemicals to the letters of the language, we have simple RNA molecules that act like genes and promote their own replication, and we have reason to think that DNA would be a better data storage device and so would ultimately replace the RNA World, where the genetic code could have plausibly originated. The conundrum at the origin of the central dogma – DNA makes RNA makes protein – is partially undone by the heroic ability of RNA to do the jobs of the others in that equation, code and function. With ribozymes, the mechanics of duplication of genes is handled by RNA itself, and doesn't need the proteins that perform this (and all) vital function. But the broader chicken-and-egg problem remains. Replication in cells from Luca onwards requires energy. So underlying the RNA World, and the transition to DNA, and the establishment of cells and life as we know it, must have been a source of energy that could be captured and manipulated and form the basis of all the energy vampire processes that cells need. And here, we find ourselves in hot, deep water.

6.   Genesis

'If you come to think of it, what a queer thing Life is! So unlike
anything else, don't you know, if you see what I mean.'

P. G. Wodehouse, *Rallying Round Old George*

It's a heavy load, but you wear it so casually. You read these words
with the accumulated weight of humankind's achievements behind
you. For all the wonder of the genetic and cellular circuits that
allow you to function, accumulated over billions of iterations in the
unbroken lineage of cells, we're also a species that is the product of
our culture. We invent, compose, trade, share, learn and create. But
more importantly, we carry these things with us through families
and through groups and across our species. Cultural evolution is
not restricted to the descent of genes through pedigrees. We can
learn from anyone else, and acquire their skills without a genetic
bond. There's a lucrative intellectual trade in declaring that one of
the things that humans do is what defines us as a species. It wasn't
the invention of tools that switched us from a simpler ancestor to
modern humans, though that was part of the process. Earlier human
species constructed simple tools, flint chippers or crude axe heads.
Plenty of animals use tools too, from otters using stones to crack
mussels to crows wielding sticks to dig out a fat grub from a log.
But what's important is not that we do uniquely human things, it is
that we continue to do them. Humans above all *accumulate* culture.

You are the sum of your parts and our species, genes passed
from parent to child for generations, and culture and ideas passed
from human to human in all sorts of directions. This should be
obvious, just as there wasn't one moment that made you who you

are. We love to pretend that there are light-bulb moments, where something flipped, and a character or an event was forged – Rosa Parks choosing to remain seated on a bus, an unnamed Chinese student defying a tank. They are iconic moments, but don't describe the flow of history. Life – your life, any life – is not merely a series of incidents. It is the accumulation of everything that you experience.

The beginning of life was just that too. The transition from chemistry to biology was the accumulation of the things that life does–feeding, copying, reproducing, and so on. At some point in the Earth's history there were just chemicals, and at a later point there was life. The most exciting question in science is how that transition happened.*

There are many components to the rebuilding of life from the bottom up, and the quest for a simple answer to a problem of immense complexity will probably be futile. How it happened the first time, and how it will happen the second, and subsequent, times, will be different. But the challenge of origin-of-life initiatives is to emulate creation in plausible and credible scenarios. That means thinking not just about what the processes were that transformed chemistry into biology, but also where.

From Foreign Shores

Why should we have such a conservative system of biology: one code, one mechanism, a central dogma? We now know that RNA probably acted as a precursor to the more robust system in place today. Francis Crick speculated that the universality of DNA was a 'frozen accident', a system that worked and was locked into place, for ever it seems, to the exclusion of any alternatives. Once in place,

---

* Even death is not easy to define, for the same reason that life isn't. We can point to events that are essential for death, those that are in opposition to life – loss of respiration, the stopped beating heart, snuffing out neural activity, the end of consciousness. But none of them is useful to the majority of life forms, which don't have brains or hearts or consciousness. Doctors use a checklist to pronounce death, but death is a process itself, not a punctuation mark at the end of a life.

any changes would either be lethal to the bearers of a novel vital system or out-competed rapidly by existing life. Why did the code become fixed? The truth is that we don't really know. Studies have worked up ideas that it was no accident, and that the four-letter code and the twenty-amino-acid lexicon are optimally balanced at a point where positive and deleterious mutation permits evolution to occur successfully. There is a seemingly simpler explanation, one which Crick was briefly seduced by. It is this: the code didn't evolve at all. It was delivered here, already functional but frozen, from somewhere else. In this case, 'somewhere else' means space.

This is one of the vaguely more scientific versions of an origin-of-life idea called 'panspermia', which says that the Earth was seeded with living things at some point in the deep past, and evolution of species ensued. Cast aside the images of space aliens, belligerent grey bipeds with spindly bodies and bulbous heads. The aliens described by panspermia's advocates would be simple, bacterial-type cells, or even simpler, the mechanics of cellular life to be harnessed by whatever the Archaean planet had available. Carried on comets or meteors, they would act as the seeds of evolution, an infection from the stars. Upon successful delivery, they were free to engage in evolution by natural selection.

Francis Crick, writing with Leslie Orgel, authored a paper in 1973 not just outlining the possibility of extraterrestrial delivery of the seeds of life, but arguing that they were sent here deliberately by intelligent beings. They were drawn to this idea – directed panspermia – largely because of the inadequacies of existing ideas about the origin of the genetic code. It's a strangely tongue-in-cheek study, contemplating out loud the limitations of space flight and speculating with wise foresight on the discoveries of planets beyond the solar system. Then, none was known; at the time of writing this we are up to almost a thousand.*

* Crick and Orgel's paper is written partially in the formal language of scientific papers, but also drifts off to wonder if the fate of the alien bio-engineers was that their star had 'frizzled them'. This is not a term that appears frequently in academic journals. It's unusual to see such august scientists pontificate in public,

It's not difficult to see the appeal of panspermia, though scientifically it is neither plausible nor credible. Even today, in this era of local galactic calm, around 30,000 tonnes of rocky debris descend onto our planet each year, mostly in the form of dust, disintegrated in the atmosphere. Showers can be seen burning up in the sky at the right time of year, such as the Perseids in each northern hemisphere summer. We even know of interplanetary relocation, at least in tiny quantities.*

Once in a blue moon, a big one hits, like in Murchison, Australia, in 1969. Mercifully rarely, a colossal one lands, like in Chicxulub sixty-five million years ago. These rocks are from space, formed when larger space rocks collide, and send their debris on a crash trajectory with Earth. Events such as this are now uncommon, especially the descent of carbon-rich meteors like the one that landed in Murchison, loaded with molecules that we know are essential for life on Earth. But leading up to our best estimates of Luca's Earth was the intense and relentless meteorite activity of the Late Heavy Bombardment. Millions of tonnes of space rocks were being deposited here over a period of millions of years, to the extent that a small proportion of the Earth's mantle is made up of interstellar rocks. It would potentially only take the survival of one single cell to seed all that followed.

Alas, the attractiveness of an idea is of no consequence in science. It's the data that counts, and therefore it's easy to deal with panspermia as an explanation for the origin of life on Earth. There is simply no evidence for it. While it is appealing in a science-fiction sense, we have no evidence for life ever having survived beyond the orbit of the moon, and there it was only us as tourists.

---

in what looks to be the formalization of a lovely pub conversation: 'The psychology of extraterrestrial societies,' they note with presumably arched eyebrows, 'is no better understood than terrestrial psychology.'

* Very occasionally, a piece of Mars lands on Earth, most recently a 7 kg meteorite called Tissint, which landed in Morocco in 2011.

That is not to say that life does not exist beyond this pale-blue dot we call home,* but we have never seen it. Not in the claims of alien abductees, nor in the microscopic bobbles of rocks from Mars,† nor in the credible scientific hunt by SETI, the Search for Extra-Terrestrial Intelligence. The rate at which we discover Earth-like planets is rocketing as we get better at looking for them. But for the time being, we are alone.

Certainly we see many of the ingredients of life in space. We are, it is important to remember, in space, formed in space and in formation part of the solar system. To ignore this is to deny the fact that our existence is determined by the space in which the Earth exists. Often, the press report a new discovery concerning the extraterrestrial origin of one of Earth's constituent parts. Invariably these studies are reported with excitable glee to be extraordinary. The true revelation of these discoveries is that they were previously unknown, not that delivery via comets or meteors might be inherently surprising. We now know that many of the ingredients of life, the component parts, are present in space, including amino acids and even, as discussed in the previous chapter, elements of genetic code. These are important scientific

---

* As Carl Sagan described it following a photograph taken by Voyager 1 in 1990, from a distance of six billion kilometres from the Earth. Many now think life elsewhere in the universe is statistically inevitable. Indeed the blossoming field of astrobiology exists to formalize the study of if, how and where life might exist off-world.

† A paper in *Science* in 1996 suggested that the shapes of particular microscopic mineral deposits on the Alan Hills meteorite (ALH 84001 in the 1990s) were biological in origin. The rock in question is thought to be around four billion years old, and was blasted off the surface of Mars in an impact. Some researchers maintain that the bobbles seen on the meteorite's surface are the product of microscopic life forms, but the scientific consensus is that they are geological in origin. Chemical analysis of the rock shows that the amino acids it bears are identical in nature to those in the Antarctic ice in which it was found, suggesting this was the very terrestrial source of biological molecules. Research continues.

discoveries, especially for the study of the origin of life. They show that the chemistry that produces biological molecules occurs outside of biology. That means that synthesis of those components is not limited to Earth. And it describes a mechanism of delivery, whereby comets or meteorites can bear these chemical parts and potentially deliver them to our surface.

The Earth has continually and unsurprisingly (though inconsistently) been the recipient of things that fill our local space. If we make a clear delineation between 'Earth' and 'not Earth', then the ingredients that make up our living planet are going to be manufactured in one of these two places. Is it a surprise that water, methane and other chemicals that figure as elemental standards in biochemistry originated off-world?

Panspermia is seductive because it looks like a simpler explanation, one that does not need the patchy, incomplete timeline that we currently have in place. But in fact it's not simpler, it's simply a dodge. If there were evidence for the transfer of life from elsewhere in the universe to Earth, and there were no other plausible explanations, then panspermia might have some legs. But it doesn't address the question of how life started on Earth, and simultaneously provides no evidential support for it starting anywhere else. Parsimony enforces the notion that it might look like a simpler explanation than our very incomplete alternatives. But it is far from being simpler because it requires not only that life abounds, or at least exists, in the universe, but that it is also based on DNA. We simply have no evidence to support that. As it is, there are other, better avenues worth exploring; they have their own problems, but at least they are scientific routes, ones that can be tested and retested. Reliance on extraterrestrial life as the delivery boy for a cosmic egg is neither necessary nor sufficient for the origin of life. For now, and for the foreseeable future, life from beyond our pale-blue dot must remain in the dominion of fiction.

Containment

All cellular processes, including the mode of information replication, are underwritten by a system that harnesses and uses energy,
and we will come to that shortly. But first, there is another issue to
deal with. Cells are discrete units of life, either in an organism or
free living, and so their information and metabolism have to be
kept separate from the rest of the universe. And so we have the
problem of the origin of the membrane. Containment of the guts
of a cell is an absolute part of being alive, and it is easy to think
that the cell membrane is merely a container, a sheer balloon in
which the machinations of life are contained. Certainly, containment of a cell's vital organs is essential as it has the effect of
concentrating chemical reactions that maintain life. But cells exist
as part of an environment, not apart from it. The cell membrane is
more akin to a customs house, an interface that monitors and controls the import and export of all the goods and messages that a
cell needs. This immense complexity allows an individual cell to
exist in its environment, or with the other cells in its host creature.
This exchange comes in the form of a highly complex network
through which traffic constantly flows, even when resting, and is
carefully regulated through gated pores, pumps and channels that
stud the surface of a cell like pips on a strawberry.

Cell membranes are built from fatty molecules called
phospholipids. These look like split pins used for holding sheaves
of hole-punched paper together: they have a water-attracting
head, and two legs that repel water. At Harvard, Jack Szostak's
work is centred on how the cellular membrane came to be. He
experiments with simpler molecules with only one tail, to see how
they behave, and how they might help us understand the formation of the first cells through self-organizing behaviour.

When Szostak mixes together a delicate blend of these fatty
molecules, they do something that is simultaneously remarkable

and not. They self-organize into a tiny bubble, about the same size as a bacterial cell, a hundredth of a millimetre wide. It's remarkable to us, because they look a bit like cells spontaneously forming out of straightforward chemicals. It's unremarkable for the reason that they do this because they can.*

The fatty acids have a bulbous head and zig-zaggy tail, and that makes them schizophrenic in their behaviour. The atoms in the head are arranged so that they like butting up against water. The tails are quite the opposite, and repel water. But the tails also attract each other. So, in the right solution, a watery one, they jostle together into two regimented lines, tails facing inwards, and heads in the water. At the right concentration, this line will expand in all directions, and link up to form a sphere. The nature of those fatty acids is such that they are chemically content when organized into this kind of membrane. As soon as that happens, there is an inside and everything else.

Modern cell membranes are studded with pumps and channels and aerials and receivers to ensure a healthy contact with the extracellular world. Biological mailboxes lie embedded in the membrane to receive input signals from around the body and local environment, and robust anchors link up neighbouring cells to hold tissue together. All of these membrane tools maintain living order within the cell and within the organism. Szostak's simple membranes are a long way from this highly evolved interface. But the first cells would not have access to that complex machinery. Part of the challenge of engineering spontaneous self-organization is to start

---

* Spontaneous self-organization is not quite as magical as it might seem. If you had Cheerios as your breakfast cereal this morning, you will have seen fundamental universal forces conspiring to invoke spontaneous organization in your bowl. The wheat rings want to float, because their density is lower than that of the milk. Gravity pushes them down, but the pressure underneath from the column of milky water beneath pushes them up. If after the first few mouthfuls there is enough space at the surface, they will automatically jiggle themselves into a hexagonal pattern, because this is the formation that allows the upwards force to be distributed evenly.

with simple molecules and evolve more complex behaviours upwards. Szostak calls his creations 'proto-cells', and many of the things they do are similar to what cells do. If you feed them, they grow, and divide. They spontaneously absorb simple genetic molecules, short fragments of things similar to DNA. When they do, this absorption triggers growth and fission. If you heat and cool them down, they not only survive intact, but DNA inside undergoes a type of replication, which can stimulate growth and division of the whole.

It's thought that these bobble-headed phospholipid molecules would not have been difficult to find on the early Earth, not because we have traces of them, but because they're easy to make with different recipes. One of the most striking is in experiments that blasted icy fragments resembling the composition of comets with ultraviolet light to mimic an interstellar forge. But they've also been found in meteorites, and made with much more Earthbound reactions. As far as ingredients go, there is ample sufficiency, and they make a good contender for the first membrane. But as the origin of life was chemistry transitioning to biology, these prototype membranes need to acquire complexity from their simple, self-organizing origin.

The membranes of modern cells need to be sophisticated to prevent leakiness. A free flow of immigration into a cell can cause all sorts of turmoil and unrest, and similarly, accidentally spilling your most prized secrets and possessions is a dangerous game. So the customs weigh-stations on the modern membrane preside over a careful coordination that will expel molecules (for messaging or as waste) and let the right ones in. It is not enough to say that the complex system we see now is simply better than what came before. In order to uncover the real transition, you have to show why one is preferential to the other. In evolution, things are never frozen and rarely accidents. So why would a complex system take over from a simpler one? Itay Budin and Jack Szostak experimentally asked that question of their proto-cells. The answer seems to come from that most Darwinian of concepts: competition.

For natural selection to play out its hand, variation is needed. If everyone was the same, not only would it be quite boring, but there would be no potential foothold for advantage. Getting one up on your siblings is what that variety produces, if you are lucky enough to bear a competitive advantage. It could be for sunlight, or food, or a mate, but without competition there is no change. Even though Darwin didn't deal with the origin of life other than idly speculating on it, not for the first time, we see a Darwinian process playing out in experiments that are most certainly pre-life. So the variation added to the stable mix of simple proto-cells in Budin and Szostak's experiment was the addition of minute quantities of complex phospholipids more like the ones in modern cells. They nestle into the membrane and sit quite comfortably next to their single-tailed colleagues. But those proto-cells grow a sixth larger than their simple brethren. So a cell that acquired the ability to make these phospholipids would immediately grow larger than one that could not. Not only this, but the increased size encourages proto-cell division.

Szostak's experiments, by his own admission, are a long way from replicating the spontaneous formation of anything as sophisticated as the contemporary cell membrane. There are all sorts of caveats to assuming that this is the re-creation of what happened the first time round. The phospholipids need to be made somehow, as they are not abundant on earth or in the heavens in the way that their single-tailed cousins are. That requires at least the most basic form of metabolism, which speculatively in an early biochemistry might be supplied by mechanisms hitherto undiscovered. On top of that, the winning cells, with their modern phospholipid membranes, lose their ability to let molecules inside and out as they please. So that demands the emergence of the channels and pores that make the modern cell membrane a customs house, rather than merely a port. Proto-cells are apparitions of cells, and do cell-like things. The simplicity of their formation is great solace for those who find the spontaneous emergence of all the facets of cellular life a big ask. Containment of living processes is necessary but not sufficient to enable life as we know it, though these experiments suggest a pathway. Nevertheless, the spontaneous formation of proto-cells in a dish is a simple yet significant step in recreating genesis.

Proto-cells also mimic cell division. In us, the process of a cell splitting in two is a highly orchestrated and active manoeuvre which requires metabolic energy. Not only does all of the DNA have to be copied, it has to be arranged for the physical split, and the membrane has to be grown big enough to make two cells from one. Modern cells have a network of internal struts and scaffold that helps everything into its right place so the split happens evenly. At the beginning of cellular life, there was presumably no such architecture. Szostak's work also demonstrates that physical properties force cell division, as simple as squeezing the proto-cells through the pore in a filter. So again, as with the experiments that show how ribozymes might have once been the original code, the formation of the cell membrane is a lot less mysterious thanks to these experiments that show how something sophisticated might have emerged spontaneously from something simpler.

Deep, Hot Water

In living cells, DNA and membranes – the information and packaging that sustain life's great descent – all require energy. The emergence of membranes and code can happen in a variety of locations, if the ingredients and conditions are right, but if we accept that, underlying these chemistries, life is a process of harnessing energy from the environment, a very different real-world scenario is required.

On a grey bench on the first floor of a building in which Charles Darwin once lived is a glass jar. It's about big enough to fit a human head in. It is perched on a wide-legged tripod, and a tin-foil-insulated rubber hose is piped into its base. Here, near-boiling water rushes in. On one side of the jar, cold water flows in through two glass nozzles. The heavy glass lid is held in place with bolts and a thick rubber seal.

This is a bioreactor. That may sound fancy, but in fact it is rather old-fashioned, like something drawn by Heath Robinson. The laboratory itself, in the aptly named Darwin Building on Gower Street in central London, is filled by University College London's phalanx of evolutionary biologists and is a typical contemporary working hive of molecular biology with benchtop spinners, shakers, machines and countless tiny tubes of colourless liquids. The bioreactor jar sits atop its tripod sprouting wires, tubes and foil, making it look somewhat ectopic in its otherwise hi-tech smart surroundings.

The magic touch is what is in the jar. About the size of an uncut Stilton, it is a block of what looks like grey stone. In fact, it's a ceramic, carefully designed to mirror the mineral growths that poke out of a newly discovered breed of submarine hydrothermal vent. It looks somewhere between a sponge and a pumice. In fact, UCL's Nick Lane, who designed this rig, experimented with both of these when trying to generate the right amount of flow-through around the material. In baking this block of sophisticated clay foam, the bubbles and pores were delicately controlled to create just the right conditions to emulate the underwater bubbling vents that Lane believes are the best candidate for the site of the first life.

The name 'vent' suggests vertical chimneys, seen in the breath-taking images of the teetering towers of the Lost City way down at the bottom of the Atlantis Massif, halfway between the Canary Islands and Bermuda. Pictures and the description of this submarine metropolis were first published at the beginning of the twenty-first century, a new type of seabed hydrothermal field, fizzing with the energy of reactions between the mantle rocks and seawater. And despite heat of up to 90 degrees Celsius and highly alkaline waters, these towers are teeming with life, so much so that lead scientist Deborah Kelly from University of Washington in Seattle told *Nature* in 2001 'you can't even see the rock because of the amount of bacteria'.

It's these vents, ever changing through the continuous effervescence from the active Earth below, that Nick Lane's ceramic is modelling. The vents emerge from the sea floor as the plates of the Earth shift and split, revealing fresh hot virgin rock drawn up from the mantle. Once exposed, they react with seawater, splitting it apart into oxygen and hydrogen and a host of other gases that are equally brimming with energetic potential, eager to react. In doing so, these gases percolate through the rocks, driving a honeycomb into them as they cool in the surrounding seawater, a process called serpentinization. It's in these tiny hatcheries that Nick Lane, Bill Martin, Mike Russell (at NASA's Jet Propulsion Lab at Caltech in California) and a small handful of other scientists think the first living processes occurred, before RNA, DNA and cell membranes. In the lab, the flow of water and gases around the porous ceramic is not simply from bottom to top, but continually circulating in and out of tiny pores, just as it does in the Lost City.

A modern cell's energy currency comes in the form of a molecule called ATP. It is a molecule that is constantly being made and recycled, as it holds energy in its chemical bonds that cells can use, to the extent that each day you will make and use your own bodyweight in ATP. The process by which this happens is a complex metabolic cycle that relies on there being a gradient of electrically charged hydrogen atoms (protons) across a membrane, a maintained imbalance. In our cells, that power generation happens in the mitochondria – the powerhouses of complex life – and in bacteria and archaea in membranes just inside their cell walls. Special proteins on these membranes act like turbines, and the flow of protons through these turbines results in the energy being generated, which gets stored in ATP, which is used to power all the processes of the cell. This cycle is so fundamental to so many living processes that it seems like a good contender for the most elemental metabolism that life can have, and therefore a system that could underlie all life. It relies simply upon there being more protons on one side of a membrane than the other – a gradient.

This is why Mike Russell has driven the idea that the hydro-thermal vents of the Lost City offer up a model for the nursery of first life. Due to the precise qualities of the chemicals bubbling up from the reactions between the rocks and the sea, they form nat-ural proton gradients in the swirls around the honeycomb rocks. In our cells that proton gradient is maintained by our biochemis-try, and it is the maintenance of that imbalance that keeps us from drifting towards decay.

The heat of the vents is not what is important: we don't use heat to generate our energy, nor indeed a bolt of lightning as Stanley Miller's experiment did, so why would first life? Cells harness chemical energy in many forms, with almost inscrutable meta-bolic pathways but with ATP produced by the flow of a proton gradient at its core. These are systems that are far from equilib-rium, and just like life, a continual restriction of increasing entropy.

In a sense, the bubbling vents are a mixture of the right ingredi-ents in such a way that biology can emerge from chemistry, but it is a far cry from the spontaneous generation of a primordial soup. In the vents it's the circulating gases that maintain the thermo-dynamic imbalance. Specifically, protons stream around the pockets in the rocks, to be concentrated into the alkaline interior away from the acid sea. So there at the bottom of the Atlantic Ocean is a bubbling bioreactor with a thermodynamic imbalance of stream-ing protons preventing retirement to equilibrium. The metal nuggets studded in the rocks catalyse the whole process, and the empty cells of the porous rock concentrate the ingredients in this never-resting mix.

As a site for the origin of life, hot vents fit seductively well, and Nick Lane's bioreactor is one of very few experiments to test that idea. It is ongoing, and at the time of writing results are unknown. If successful, Lane hopes to see the emergence of reactions and chemicals that are similar to or the same as the ones we see in biol-ogy. This might then suggest a place for Luca's birth.

Bill Martin pushes the model even further. To him Luca is a thing which is not a free-living cell at all.★ Even with a bushy network between single cells that populated the Earth for two billion years, Luca still stands as a common ancestor of both archaea and bacteria. In Martin's version, Luca was a swill of activity, but not a free-living cell as we might have assumed. The beginnings of life were not inside a membrane-bound cell. Instead, the last common ancestor of the two oldest domains of life was locked inside the rocky shell of the alkaline underwater vents. In there, concentrated and protected, but fed and with a constant supply of charged atoms, is the transition from chemistry to biochemistry, and then on to life. The evolutionary split that wrenches archaea from bacteria happens after the RNA World has given way to DNA, but before the arrangement of fatty molecules into the cell membrane. They have the same genetic code, and they both rely on ribosomes as their protein factories. In both, they are more similar to each other than to our cells. But in archaea, the tools for writing and copying RNA – a protein called RNA polymerase – are more similar to our own than they are to bacteria's. The archaea's outermost shell, the cell wall, is different from a typical bacterial cell wall, and the membrane is radically different from anything else. These differences are present in contemporary species, but they are fundamental enough to have roots beneath the bottom of the intertwined bushy bit of the tree of life. According to Martin, first life was housed not in membranes, but in rock.

★ This is a heterodox view from a man who likes a healthy bit of agitprop. He starts his lectures on the origin of life by tipping up the soup bowl. He points out that Haldane and Oparin, the independent founders of the soupist model, were both Communists. Coming from a Texan (who now lives in Germany), this instantly sounds like a political point. In fact, he is quick to point out, it means that, as they were both dialectic materialists, 'there had to be a rational explanation for everything from the workings of society to the needs of individuals to the origin of life. And [Haldane and] Oparin's theory became party doctrine.'

Martin's model starts with bubbling hydrogen, ammonia and hydrogen sulphide frothing around the pores in a serpentinized rock. These pockets look rather like the empty cork chambers that Robert Hooke saw with his microscope at the very beginning of this journey – not cells as we know them, but cells like the honeycomb cavities of a beehive. The acid sea and the alkali interior provide a natural proton gradient for the flow of power that incepts a basic form of metabolism, and it is continuous for as long as the vent is active. Simple biochemical reactions that we observe in modern cells begin to occur in this energetic mix, and we start to see amino acids being forged in this tumult. Next come other foundation biomolecules, such as sugars, purine and pyramidine rings and other molecules that we see in metabolic cycles. Purine and pyramidine go on to fuse, becoming bases – the letters of genetic code, maybe as they do in John Sutherland's syntheses. When these letters link, the RNA World can begin, only to be eventually replaced by DNA with its superior data-storage prowess. But it's all still trapped in the tortuous labyrinths of the rock, the pores concentrating biochemical action, rather than the dilution of a warm pond of a broth. This locked reactor is the last common ancestor of everything, rock-bound for now, but soon acquiring the skin that will allow it to break free.

It's here that life's first great schism occurs. Two sets of molecules grow to line the inside of separate rocky chambers that hold this brimming biochemistry. One will lead to bacteria, and the second, to every other living thing. The biochemistry that has developed in these cells is shared, but from now on, as they evolve new powers, they build different types of membrane to house themselves, and new ways of maintaining the energetic difference between outside and in. There, stuck to the side of a gassy rock at the bottom of the ancient sea, cellular life begins.

This is an idea. It is Bill Martin's plausible description of how genesis could have occurred based on what we know about simple life and the chemistry and geology of the vents. It's impossible to put a timescale on this transition. Roughly, we have a window of several hundred million years, somewhere between the Earth-pummelling Late Heavy Bombardment and the fossil cells we can actually see, around 3.6 billion years old. Modern alkali vents are fairly stable, but rare. They may have abounded on the sea floors of the capricious young Earth. Chemical reactions tend not to dawdle. Allowing a million years for two things to react gives ample opportunity for them not to, to be washed away. But this timescale provides the opportunity for the experiment to take place billions of times, over and over again in billions of pores, with infinite variables being tested over and over again. It is a conspiracy of chance made possible by the sheer numbers on offer. It only needs to have worked once, the jackpot in the chemical lottery.

We can't repeat the timescale, nor can we experiment in vents themselves. But we can model them. This is why Nick Lane's humble bioreactor is such an important experiment. It is an experiment in the purest sense too, as he really doesn't know what will come out of it. The pores are seeded with iron sulphide, fool's gold, a catalyst that is present in the vents, and may speed up the energy transduction that is the beginning of metabolism. Just like in the vents, protons bubble up from the base, and circulate through the labyrinthine pores. Each of many different biological processes is being tested individually, and the products examined. Lane is looking for the hallmarks of biological molecules – the bases of RNA, and their component parts, amino acids that make up proteins, and the molecules that come from energy-harnessing metabolism. It is designed to find hints of biology built without enzymes, and at the time of writing, this simulation is just beginning. The key to this experiment is that it is all far from equilibrium, just like life is, and just like the vents are. It won't make cells, but it might spontaneously generate signatures of the chemistry that underlies all living processes.

Although the definition might be elusive, we certainly know

what life is when we see it. Life is physics, chemistry and biology. That makes the attempts to re-create its origins a renaissance science. That also makes it contentious, because different experts will approach from the angle that makes most sense to them. The quest to build life anew is to understand all of the things that life does, and to re-create them individually at first, then gradually align them and fuse them. Many of the pieces are being forged, all with different but vital characteristics: energy, information, reproduction, metabolism, evolution.

The journey from simple chemistry, to more complex chemistry, to biochemistry, from simple information to profoundly inscrutably sophisticated genetics, and from simple soapy bubbles to dynamic customs houses of membranes might seem unlikely. The astronomer and author Fred Hoyle once described the unlikelihood of the animation of chemistry as follows: 'The chance that higher life forms might have emerged in this way is comparable to the chance that a tornado sweeping through a junkyard might assemble a Boeing 747 from the materials therein.' He goes on, in typically bombastic terms, to say that it is 'nonsense of a high order'. Writing in his book *Evolution from Space*, he made a calculation about the probability of the proteins of a most basic living cell arising spontaneously, and came up with the number 1 with 40,000 zeroes after it.* Hoyle believed in panspermia and used this calculation to dispute the origin of life being the earthbound transition from chemistry to biology. It was a fallacious argument, and has been refuted sensibly many times. Hoyle's errors include the assumption that modern proteins were the first in life. He assumes that the trial of building a working cell was sequential, and not simultaneous. He may have been contemplating the twenty amino acids that emerged from the sludge of Stanley Miller's famous

* Hoyle was a brilliant scientist and author, but also something of an iconoclast, prone to vocally rejecting mainstream ideas. He disputed the universe's origin being the result of the Big Bang, which is the overwhelming scientific consensus view.

brew, all present and correct, ready to fall into line into a working, living entity.

But in the works of Jerry Joyce, Jack Szostak, John Sutherland, Nick Lane, Mike Russell, Bill Martin and others we now see emergent properties that are not improbable at all, and are a lot less mysterious than we once thought. The conditions of the infant Earth, tested in modern labs, render self-creation unavoidable.

We see RNA whose origin is random sequence fulfilling the roles of early enzymes and early genes. We see cell-like bags summoned into being by nothing other than atomic forces. And we experiment to see the hallmarks of metabolism emerge from the effervescence of inanimate chemistry. As we march towards an increasingly sturdy model of genesis, the mysteries of creation once filled by deities inevitably give way to science. In the lab at least, we see a foreshadowing of the process that has driven the spread of life for four billion years. Ribozymes undergo a form of Darwinian selection, as does the transition from simple membranes to complex. This is not to say that Darwin had miraculous prescient knowledge of chemistry a hundred years on from his death. It's merely that the process of natural selection that he first described is more powerful than he could have possibly imagined. And 'whilst this planet has gone cycling on according to the fixed law of gravity', as he wrote, a transition happened, driven by the power of a process that would one day lead directly to you. As we continue to explore the world of the cell, we are on the brink of seeing the start of this grand journey of life on Earth, the only living planet we know of. And such humble beginnings. Every muscle twitch, every breath, every thought, emotion and sensation you've ever had, right down to the wince of pain from an insignificant cut, started its journey in a microscopic chamber at the bottom of the sea, four million millennia ago.

References, Notes and Further Reading

In this book I refer to a single bacterial cell as 'a bacteria' rather than 'a bacterium', on the principle that its adoption into the English language releases it from the grammatical rules of its language of origin.

We luxuriate in a time where the basic mechanisms of life on Earth are largely understood, and many great books have been written that describe evolution and genetics. My recent favourites include *Life Ascending* by Nick Lane (Profile Books, 2010), *Why Evolution Is True* by Jerry Coyne (Oxford University Press, 2010), *Your Inner Fish* by Neil Shubin (Penguin, 2009), many of the books of Matt Ridley and almost everything by Steve Jones, particularly *The Language of the Genes* (HarperCollins, 1993, which, despite being written twenty years ago, and crucially ten years before the Human Genome Project was first published, contains insight and stories as good as any contemporary book on the nature of DNA), and *Almost Like a Whale* (Black Swan, 2000), which successfully updates the *Origin of Species* with contemporary evolutionary biology. Of course, reading anything by Charles Darwin will make you cleverer. His works are as important to our culture as anything by any writer.

Chapter 1: Begotten, Not Created

As with all science, the pathways to great discoveries are complex, tortuous and winding, and almost always bereft of eureka moments. The road that led from van Leuwenhoek to cell theory is no different, with dozens of men making small incremental experimental and conceptual advances. I have outlined the narrative of just some of the key players. For a thorough and scholarly history of the early microscopists, Professor Sir Henry Harris' book *The Birth of the Cell* (Yale University Press, 2000) is the definitive guide, from which I have sourced many sections in the story of cell theory.

Animalia by Aristotle, 350 BCE, translated by D'Arcy Wentworth
    Thompson in 1910, is thoroughly readable and well worth a look, and
    is available free online at http://classics.mit.edu/Aristotle/history_
    anim.html.

Chapter 2: Into One

Plenty has been written about Gregor Mendel, as befits one of the great
    scientists, and here is the original paper where he outlines what would
    become the laws of inheritance: G. Mendel, 'Versuche über Pflanzen-
    hybriden', *Verhandlungen des Naturforschenden Vereines* 4, Abh. Brünn
    (1866), pp. 3–47, or in translation: 'Experiments in Plant Hybridiza-
    tion', *Journal of the Royal Horticultural Society* 26 (1901), pp. 1–32.
That there were 115 copies of that paper, none of which made it into Dar-
    win's library, is documented here: R. C. Olby, 'Mendels Vorlaüfer:
    Kölreuter, Wichura, und Gärtner', *Folia Mendeliana* 21 (1986), pp. 49–67.
One point of Mendelian interest lies in re-analysis of his data by Ronald
    Fisher, one of the founders of modern evolutionary biology (along
    with J. B. S. Haldane, whom I'm told he could not stand to be in the
    same room as, and Sewall Wright). Fisher concluded that the statistical
    significance of Mendel's results meant that the experiments must have
    been 'falsified so as to agree closely with Mendel's expectations'. He
    wasn't alleging that the laws are incorrect, merely that the results were
    precise enough to suggest massaging. This has been repeated often
    since, but the following paper by the geneticists Daniel Hartl and Dan-
    iel Fairbanks concludes that this allegation can be put down 'because on
    closer analysis it has proved to be unsupported by convincing evidence':
    Daniel L. Hartl and Daniel J. Fairbanks, 'Mud Sticks: On the Alleged
    Falsification of Mendel's Data', *Genetics* 175:3 (March 2007), pp. 975–9.
The story of twentieth-century biology and indeed science orbits
    around probably the most significant scientific advance of that
    century – Crick and Watson's publication of the double helix. For
    more on the story of DNA, there are plenty of books available. James
    Watson's own account, *The Double Helix*, is a gripping read (reissued
    by Simon and Schuster, 2012), though it is very clearly a personal
    account by a fabulous storyteller, and to my mind bears the signature
    of myth-making. His treatment of Rosalind Franklin in that book is
    horrid. Franklin's role in the story of DNA is frequently debated,

often with the question 'Should she have got the Nobel Prize alongside Crick and Watson?' There is a simple and absolute answer to
that, which is no: the rules are that Nobels will not be given post-
humously, and Franklin was prematurely dead by the time DNA
was recognized by the awarding committee in 1962. Brenda Mad-
dox's biography *Rosalind Franklin: The Dark Lady of DNA*
(HarperCollins, 2003) is thrilling, comprehensive and honest, por-
traying Franklin in a believable and not always flattering light. It is a
marvellous book.

Francis Crick, as befits his genius, appears in several different sections of
this book, as he continued to explore the nature of life after Watson
and his grand discovery in 1953. There isn't a better description of his
life and works than Matt Ridley's *Francis Crick: Discoverer of the Genetic
Code* (HarperCollins, 2011).

The BBC dramatization of the tale of DNA is called *Life Story* (1987),
and is thoroughly watchable. It features Hollywood's favourite
otherworldly quirky scientist actor, Jeff Goldblum, in the role of
otherworldly quirky scientist Jim Watson. As an aside, Goldblum has
also played a neurosurgeon (*The Adventures of Buckaroo Banzai Across
the 8th Dimension*, 1984), a particle physicist (*The Fly*, 1986), a mathem-
atician (*Jurassic Park*, 1993), a computer scientist and environmentalist
(*Independence Day*, 1996), a marine biologist (*The Life Aquatic*, 2004),
and a canine immunologist (*Cats and Dogs*, 2001). Can any other actor
boast such a litany of scientists?

'We wish to suggest a structure for the salt of deoxyribonucleic acid
(D.N.A.). This structure has novel features which are of considerable
biological interest.' So begins probably the most famous and significant research paper of the twentieth century: J. D. Watson and F. H. C.
Crick, 'A Structure for Deoxyribose Nucleic Acid', *Nature* 171 (25 April
1953), pp. 737–8. This is available at www.nature.com/nature/dna50/
archive.html, along with several other landmark papers by titans of
biology from this golden age, including: O. T. Avery, C. M. MacLeod
and M. McCarty, 'Studies on the Chemical Nature of the Substance
Inducing Transformation of Pneumococcal Types', *Journal of Experimental Medicine* 79 (1944), pp. 137–59.

Language is the most useful and frequently used metaphor for understanding and explaining the workings of DNA. Language also shares
many similar evolutionary characteristics, as explained in this
review by biologist Mark Pagel: 'Human Language as a Culturally Transmitted Replicator', *Nature Reviews Genetics* 10 (2009),

pp. 405–15 (doi:10.1038/nrg2560). And for a terrifically enjoyable non-scientific romp though the evolution of words and language read: Guy Deutscher, *The Unfolding of Language: The Evolution of Mankind's Greatest Invention* (Arrow, 2006).

For Douglas Theobald's statistical analysis of Luca's existence see: Douglas L. Theobald, 'A Formal Test of the Theory of Universal Common Ancestry', *Nature* 465 (13 May 2010), pp. 219–22 (doi:10.1038/nature09014); and a rebuttal: Takahiro Yonezawa and Masami Hasegawa, 'Was the Universal Common Ancestry Proved?', *Nature* 468, E9 (16 December 2010), pp. 219–22 (doi:10.1038/nature09482).

Laurence D. Hurst and Alexa R. Merchant, 'High Guanine-cytosine Content Is Not an Adaptation to High Temperature: A Comparative Analysis Amongst Prokaryotes', *Proceedings of the Royal Society B* 268 (2001), 493–7 (doi:10.1098/rspb.2000.1397).

On some of the earliest fossils of cells: David Wacey et al., 'Microfossils of Sulphur-metabolizing Cells in 3.4-billion-year-old Rocks of Western Australia', *Nature Geoscience* 4 (2011), pp. 698–702 (doi:10.1038/ngeo1238).

There are many papers on bacteria and archaea swapping genes rather than inheriting them from their cellular parents. One of the most graphic is: V. Kunin et al., 'The Net of Life: Reconstructing the Microbial Phylogenetic Network', *Genome Research* 15 (2005), pp. 954–9, which features a necessarily baffling diagram of the tangled bank of the base of the tree of life. *New Scientist*, a magazine with a proud history, decided to focus their attention to this precise subject on the 200th anniversary of Darwin's birth in 2009. The bold cover line was 'Darwin Was Wrong', though the articles inside were more nuanced about the subject of lateral or horizontal gene transfer. They explained in an accompanying editorial:

As we celebrate the 200th anniversary of Darwin's birth, we await a third revolution that will see biology changed and strengthened. None of this should give succour to creationists, whose blinkered universe is doubtless already buzzing with the news that '*New Scientist* has announced Darwin was wrong'. Expect to find excerpts ripped out of context and presented as evidence that biologists are deserting the theory of evolution en masse. They are not.

On seven separate occasions, I was shown this cover (or someone quoted it) at public meetings as part of attacks from particular religious groups who choose to ignore evidence and assert creationist doctrine.

Chapter 3: Hell on Earth

See Euan Nisbet's classic textbook on the first couple of billion years of our planet: E. G. Nisbet, *The Young Earth: An Introduction to Archaean Geology* (Allen and Unwin, 1987).

A study suggesting that comets delivered water to the Earth and the moon is: James P. Greenwood et al., 'Hydrogen Isotope Ratios in Lunar Rocks Indicate Delivery of Cometary Water to the Moon', *Nature Geoscience* 4 (2011), pp. 79–82 (doi:10.1038/ngeo1050).

An interesting model that suggests that microbial life could have survived underwater the intense barrage of the Late Heavy Bombardment can be found in: Oleg Abramov and Stephen J. Mojzsis, 'Microbial Habitability of the Hadean Earth During the Late Heavy Bombardment', *Nature* 459 (21 May 2009), pp. 419–22 (doi:10.1038/nature08015); Lynn J. Rothschild, 'Earth Science: Life Battered but Unbowed', *Nature* 459 (21 May 2009), pp. 335–6 (doi:10.1038/459335a).

Simon A. Wilde et al., 'Evidence from Detrital Zircons for the Existence of Continental Crust and Oceans on the Earth 4.4 Gyr Ago', *Nature* 409 (11 January 2001), pp. 175–8 (doi:10.1038/35051550).

One of the key studies that showed that the Moon was blasted off the very young Earth by an impact from another large celestial body, Theia, the Titan of ancient Greek myth who gave birth to the Moon goddess, Selene, is: U. Wiechert et al., 'Oxygen Isotopes and the Moon-Forming Giant Impact', *Science* 294 (12 October 2001), pp. 345–8 (doi:10.1126/science.1063037).

On Darwin's handwriting: as part of the filming of BBC4's *The Cell*, I had the great privilege to hold in my ungloved hands the letter that Darwin sent to Joseph Hooker in 1871 in which he ponders his now-famous 'warm little pond'. Standing in a reading room in the University of Cambridge Library, I read an extract from this priceless piece of history. But after several takes spoilt because I couldn't decipher his scrawl, we ended up slipping a printed transcript on top of the actual letter. It is, of course, entirely irrelevant that his handwriting was so awful, but I find this quite amusing.

J. B. S. (Jack) Haldane is one of the dominant figures of twentieth-century biology in many of its forms, and a truly fascinating chap. Not only did he ponder the origin and nature of life and species, he was a central figure in the fusion of genetics and evolutionary biology and many other fundamental aspects of the living world, and was probably the first person to suggest hydrogen as a basis for renewable energy (in 1923). He was a Marxist, and vocally spoke out against the British Government's role in the Suez crisis in 1956. His classic essay 'On Being the Right Size' is freely available online, as is a very curious film presented by Haldane (produced by the Soviet Film Agency, 1940) entitled *Experiments in the Revival of Organisms*, which features the posthumous physical reactions of a severed dog's head. The definitive biography by Ronald Clark, *The Life and Works of J. B. S. Haldane* (1968), unfortunately is currently out of print.

This is Stanley Miller's iconic experiment from 1953: S. L. Miller, 'A Production of Amino Acids under Possible Primitive Earth Conditions', *Science* 117 (15 May 1953), pp. 528–9 (doi:10.1126/science.117.3046.528).

And Jeffrey Bada's 2008 analysis of some of Miller's experiments: Adam P. Johnson et al., 'The Miller Volcanic Spark Discharge Experiment', *Science* 322 (17 October 2008), p. 404 (doi:10.1126/science.1161527).

Chapter 4: What Is Life?

A good read on the accumulation of information storage and inheritance in molecules is: G. F. Joyce, 'Bit by Bit: The Darwinian Basis of Life', *PLoS Biology* 10 (2012) (e1001323. doi:10.1371/).

About various definitions of life: Pier Luigi Luisi, *Origins of Life and Evolution of the Biosphere* 28 (1998), pp. 613–22.

Edward Trifonov's meta-definition of life based on the words that scientists use. Nineteen retorts can be found in the February 2012 edition of the same journal: Edward N. Trifonov, 'Vocabulary of Definitions of Life Suggests a Definition', *Journal of Biomolecular Structure and Dynamics* 29 (2011), pp. 259–66.

Both Erwin Schrödinger's and J. B. S. Haldane's pamphlets entitled 'What Is Life?' (1944 and 1949, respectively) are available online for free, and are both essential reading.

Chapter 5: The Origin of the Code

Kevin Leu et al., 'On the Prebiotic Evolutionary Advantage of Transferring Genetic Information from RNA to DNA', *Nucleic Acids Research* 39 (2011), pp. 8135–47 (doi:10.1093/nar/gkr525).

On Jerry Joyce and Tracey Lincoln's short RNA molecules that have the effect of endlessly reproducing themselves: Tracey A. Lincoln and Gerald F. Joyce, 'Self-Sustained Replication of an RNA Enzyme', *Science* 32 (27 February 2009), pp. 1229–32 (doi:10.1126/science.1167856).

On Jack Szostak and David Bartel's classic ribozyme experiment: D. P. Bartel and J. W. Szostak, 'Isolation of New Ribozymes from a Large Pool of Random Sequences', *Science* 261 (10 September 1993), pp. 1411–18.

David Adam, 'Give Six Monkeys a Computer, and What Do You Get? Certainly Not the Bard', *Guardian*, Friday 9 May 2003.

On the idea of a ribozyme being the most distant ancestor inspired by the blog Tales from the Nobel Factory, written by Alex Taylor: though I have chosen a different ribozyme as a conceptual aboriginal ancestor, I am most grateful to him for this and other bits he helped me with, though not his inability to understand what a collective noun is: http:// talesfromthenobelfactory.posterous.com/; Aniela Wochner et al., 'Ribozyme-Catalyzed Transcription of an Active Ribozyme', *Science* 332 (8 April 2011), pp. 209–12 (doi:10.1126/science.1200752).

Three papers on the origin of the genetic code with fewer than the four bases modern life bears: Jeff Rogers and Gerald F. Joyce, 'A Ribozyme That Lacks Cytidine', *Nature* 402 (18 November 1999), pp. 323–5 (doi:10.1038/46335); John S. Reader and Gerald F. Joyce, 'A Ribozyme Composed of Only Two Different Nucleotides', *Nature* 420 (19 December 2002), pp. 841–4 (doi:10.1038/nature01185); Julia Derr et al., 'Prebiotically Plausible Mechanisms Increase Compositional Diversity of Nucleic Acid Sequences', *Nucleic Acids Research* (2012) (doi:10.1093/nar/gks065).

John Sutherland's classic synthesis of uracil: Matthew W. Powner, Béatrice Gerland and John D. Sutherland, 'Synthesis of Activated Pyrimidine Ribonucleotides in Prebiotically Plausible Conditions', *Nature* 459 (14 May 2009), pp. 239–42 (doi:10.1038/nature08013).

On uracil from space: Zita Martins et al., 'Extraterrestrial Nucleobases in the Murchison Meteorite', *Earth and Planetary Science Letters* 270 (2008), pp. 130–36.

Chapter 6: Genesis

On the accumulation of human culture: Adam Powell, Stephen Shennan and Mark G. Thomas, 'Late Pleistocene Demography and the

Appearance of Modern Human Behavior', *Science* 324 (5 June 2009), pp. 1298–1301 (doi:10.1126/science.1170165).

On competition in cell membrane formation: Itay Budin and Jack W. Szostak, 'Physical Effects Underlying the Transition from Primitive to Modern Cell Membranes', *PNAS* (2011) (doi:10.1073/pnas.1100498108).

Jack Szostak's proto-cell model dominates studies of the emergence of the membrane. Other ideas and experiments have suggested different routes. One recent one by Stephen Mann, Shoga Koga and colleagues from the University of Bristol suggests microdroplets stuffed with nucleotides and amino acids strings so that they compartmentalize without a physical skin. We await the development of that idea with relish. Shogo Koga et al., 'Peptide-nucleotide Microdroplets as a Step Towards a Membrane-free Protocell Model', *Nature Chemistry* 3 (2011), pp. 720–24 (doi:10.1038/nchem.1110).

Here's an interesting study from 2011 that shows replication of DNA within self-reproducing proto-cells and sharing of the DNA between the daughter cells. Although very clearly not the same as in modern cells, the process mimics the outcome of actual cell division: K. Kurihara et al., 'Self-reproduction of Supramolecular Giant Vesicles Combined with the Amplification of Encapsulated DNA', *Nature Chemistry* 3 (2011), pp. 775–81 (doi:10.1038/nchem.1127).

Leslie E. Orgel, 'Prebiotic Chemistry and the Origin of the RNA World', *Critical Reviews in Biochemistry and Molecular Biology* 39 (2004), pp. 99–123 (doi:10.1080/10409230490460765).

Jack W. Szostak, David P. Bartel and P. Luigi Luisi, 'Synthesizing Life', *Nature* 409 (18 January 2001), pp. 387–90 (doi:10.1038/35053176).

The first description of the hydrothermal vents of the Lost City: D. S. Kelly et al., 'An Off-axis Hydrothermal-vent Field Near the Mid-Atlantic Ridge at 30° N', *Nature* 412 (12 July 2001), pp. 145–9.

There are a handful of papers and reviews by Nick Lane, Mike Russell, Bill Martin and others on proton gradients and the energetic origin of life in hydrothermal vents. Here are some of the best: N. Lane and W. Martin, 'The Origin of Membrane Bioenergetics', *Cell* 151 (2012), pp. 1406–16; W. Martin, 'Hydrogen, Metals, Bifurcating Electrons, and Proton Gradients: The Early Evolution of Biological Energy Conservation', *FEBS Letters* 586 (2012), 485–93; M. J. Russell (ed.), *Origins, Abiogenesis and the Search for Life* (Cosmology Science Publishers, 2011); N. Lane, J. F. Allen and W. Martin, 'How Did LUCA Make a Living? Chemiosmosis in the Origin of Life', *Bioessays* 32 (2010), pp. 271–80; W. Martin, J. Baross, D. Kelley, M. J. Russell,

'Hydrothermal Vents and the Origin of Life', *Nature Reviews, Microbiology* 6 (2008), pp. 806–14; W. Martin and M. J. Russell, 'On the Origin of Biochemistry at an Alkaline Hydrothermal Vent', *Philosophical Transactions, Royal Society of London* (Ser. B) 362 (2007), pp. 1887–925; and this is a profile of Mike Russell and his research: John Whitfield, 'Origin of Life: Nascence Man', *Nature* 459 (2009), pp. 316–19 (doi:10.1038/459316a).

If you can get hold of this paper, it is a hoot: Crick and Orgel pondering aliens as the origin of earthly life: F. H. C. Crick and L. E. Orgel, 'Directed Panspermia', *Icarus* 19 (1973), pp. 341–6.

Sir Fred Hoyle and Chandra Wickramasinghe's interesting but wrong ideas on panspermia: Sir Fred Hoyle and Chandra Wickramasinghe, *Evolution from Space: A Theory of Cosmic Creationism* (Touchstone, 1984).

Index

Creation

Creation

The Future of Life

ADAM RUTHERFORD

VIKING
an imprint of
PENGUIN BOOKS

VIKING

Published by the Penguin Group

Penguin Books Ltd, 80 Strand, London WC2R ORL, England

Penguin Group (USA) Inc., 375 Hudson Street, New York, New York 10014, USA

Penguin Group (Canada), 90 Eglinton Avenue East, Suite 700, Toronto, Ontario, Canada M4P 2Y3
(a division of Pearson Penguin Canada Inc.)

Penguin Ireland, 25 St Stephen's Green, Dublin 2, Ireland (a division of Penguin Books Ltd)

Penguin Group (Australia), 707 Collins Street, Melbourne, Victoria 3008, Australia
(a division of Pearson Australia Group Pty Ltd)

Penguin Books India Pvt Ltd, 11 Community Centre, Panchsheel Park, New Delhi – 110 017, India

Penguin Group (NZ), 67 Apollo Drive, Rosedale, Auckland 0632, New Zealand
(a division of Pearson New Zealand Ltd)

Penguin Books (South Africa) (Pty) Ltd, Block D, Rosebank Office Park,
181 Jan Smuts Avenue, Parktown North, Gauteng 2193, South Africa

Penguin Books Ltd, Registered Offices: 80 Strand, London WC2R ORL, England

www.penguin.com

First published 2013
001

Copyright © Adam Rutherford, 2013

The moral right of the author has been asserted

Set in 12/14.75 pt Bembo Book MT Std
Typeset by Jouve (UK), Milton Keynes
Printed in Great Britain by Clays Ltd, St Ives plc

A CIP catalogue record for this book is available from the British Library

HARDBACK ISBN: 978-0-670-92044-0
TRADE PAPERBACK ISBN: 978-0-670-92046-4

www.greenpenguin.co.uk

ALWAYS LEARNING   **PEARSON**

To Beatrice and Jake, who came from my cells

Contents

Foreword

The best books about evolution have probably already been written. That is testament to the grand, brilliant and demonstrably correct idea that underlies it. In November 1859, Charles Darwin published *On the Origin of Species*, in which he outlined his thoughtful arguments about evolution via the mechanism of natural selection. Although the progress of science dictates that theories and models are constantly evolving, in the 150 years since that publication, every aspect of biological research has effectively served to reinforce the central idea in the *Origin of Species*, that species change over time with the gain or loss of characteristics according to their usefulness – often referred to with Darwin's own aphorism: 'descent with modification'. A century and a half of research has shown beyond reasonable doubt quite how robust an idea natural selection is.

This book is about what came before the origin of species, and what comes next, as we engineer life forms outside of the constraints of natural selection. In Hollywood terms, it concerns the prequel and the sequel to life. Both halves are also concerned with modified descent. In this part we will explore the modification of life by human agency – the designing, engineering and building of new life forms for purpose. The other part traces the search for the origin of life – the ascent of lifeless chemistry into biology, in the tumult of the elemental rocks, seas and the bubbling swirl of the infant Earth. To explore these endeavours, we need to understand four billion years of evolution, and the two or three centuries of biology that have led to this critical point in human history. These two fields are intimately interdependent, a tangled thicket of ideas whose connections are testament to the scientific method and our insatiable curiosity to find things out. As we have come to

an ever-greater understanding of the processes at the beginning of evolution, we've learned how to profoundly manipulate biology in the present, and vice versa: as we take cells apart and reassemble them synthetically we learn more about the conditions in which the first cells arose.

For that reason, this is a book in two halves, each of which can be read independently, though both stories are inextricably connected. There are two front covers, and this foreword, modified with minor suitable adaptations, appears at the beginning of both. To begin with the origin of life, turn the book upside down now.

The Future of Life

Introduction

'What I cannot create, I do not understand.'

Richard Feynman, 1988

What is the best way to find out how a thing works? There are a number of approaches, including observing how it behaves, or testing it (possibly to destruction or at least till it fails). But there is only so much you can learn from these techniques. To understand a thing of complexity you have to take it apart. A car mechanic cannot understand the complexities of the internal combustion engine merely by thrashing it on a racetrack. To understand an engine you must take it apart and see how the parts fit together, what each one does separately and how they relate to one another.

Since before the discovery of the most basic unit of life in the late seventeenth century, the cell,★ biologists have spent plenty of their time observing and testing how living things work. But a much more instructive endeavour has been to take them apart. Leonardo da Vinci was an accomplished anatomist, who drew many beautifully intricate diagrams of the insides of animals and people, to determine structures and functions within our bodies. In the early seventeenth century William Harvey deduced the

★ Some scientists consider viruses, or at least some types of virus, to be living. They are problematic as they have many characteristics of the more definably living cells, but lack the cellular machinery to reproduce themselves. Thus they are parasitic on living cells in order to reproduce. The question 'What is life?' is explored in Chapter 4 of the other half of *Creation*, but for simplicity's sake I have chosen to adhere to the general, though not undisputed, consensus that viruses are not alive.

mechanisms of blood circulation by dissecting the fine construction of veins, arteries and the animal and human heart.

Fifty years later, the Dutch draper Antonie van Leuwenhoek
had discovered cells, the things from which all life forms are built.
As microscopes improved, the guts of these tiny bags of living
matter were systematically unveiled, and in time removed from
the protective membranes that encase their innards. In the nineteenth century, mere observation was giving way to chemical
detective work, and the composition of the components of cells –
proteins, and other essential ingredients of life – were being
described.

It was during that century with these chemical techniques that
DNA was first isolated, though it was almost 100 years before its
importance and its iconic double-helix structure were discovered.
In 1869, a young doctor called Friedrich Miescher was working in
Tübingen in Germany. As a result of the ongoing Franco-Prussian
war, he had ready access to the pus-soaked bandages of wounded
soldiers slowly rotting in a local hospital. Pus contains a surfeit of
white blood cells called leucocytes, their purpose being to fight
the determined invasion of disease-causing interlopers at an open
wound. In extracting various ingredients from the fetid bandages,
he isolated a chemical that contained significant quantities of the
small molecular group phosphate. This particular cluster of five
atoms is not normally detected in proteins, whose chemical composition had been well studied by this time. So Miescher figured
his extract was different. He called it 'nuclein', as it came primarily
from the nucleus of the soldiers' leucocytes, a separate compart-
ment in the centre of the cells of all complex life. Protein names
tend to end in the letters '-in', such as haemaglobin or insulin, so it
is tempting to speculate that Miescher was still thinking of it in
similar terms. His work was never really followed up, and it wasn't
until many years later, when nuclein was shown to belong to an
entirely different class of molecule, that it was identified as DNA.

In the twentieth century biology matured into the study of
ever-smaller components of cells. DNA was characterized

famously by Watson and Crick using the data of Rosalind Franklin and Maurice Wilkins in 1953, and Crick, along with dozens of others, spent the next few years decrypting how vital information was stored inside the double helix. These biological data are the instructions to make an organism, and each species has its own set, known as a 'genome'. How it works for us was the focus of the grandest endeavour in biology, the Human Genome Project. That venture finished in the first decade of the twenty-first century with a complete read-through of the three billion letters of DNA in an average human, though the mysteries of human genomics continue to be explored.

From Crick and Watson's landmark on, the second half of the twentieth century was the era of molecular biology, and by the end of it the majority of all scientific research on Earth broadly concerned the molecules of life – DNA, its equally important cousin RNA and proteins. Understanding the molecules and the language of life and their universal nature has profoundly transformed all aspects of our understanding of living organisms. As we explore in the other half of this book, DNA and molecular biology were the final bricks that cemented our understanding of the nature of evolution and provided a unifying mechanism by which we can trace our way back into the deep past, and to a single origin of life, a stem from which all life has grown.

But it has also spawned a new way to examine living things. As we will discover in the following chapters, our robust understanding of the molecules of life has led to an era in which we can alter, manipulate and effectively remix the basic genetic code of any living thing. We collect these related fields under the umbrella term of 'genetic engineering', which has already begun to transform aspects of our lives. It has revolutionized the discovery of how diseases transpire, as we can alter the DNA of animals and cells to mimic those diseases and use those remixed cells as a test ground for audacious new medical treatments.

If the biological twentieth century was concerned with taking cells apart to understand how they work, this newfound

understanding has also given us the ability to put them back together again, but intelligently designed by us with specific purposes. Cells are minute manufactories, evolved over billions of years to have highly specialized functions – to construct bone, or store memories, convert light into electricity in our eyes, or swallow invaders that cause us harm. They either are collected into a community performing in harmony to form an organism or, as is the case with most cells on Earth (bacteria and their lesser-known cousins archaea), live free as single entities.

The quotation at the beginning of this introduction, 'What I cannot create, I do not understand,' is attributed to the great bongo-playing physicist and all-around master of science and science communication Richard Feynman. It was his last message to his students, written on his blackboard at Caltech at the time of his death in 1988. It is in the reassembly of the language of life that we are at a place beyond mere comprehension of living systems, one where life forms are engineered tools. Genetic tinkering became wholesale engineering, and now has evolved into a new field whose purpose is solely to create life forms that serve as tools for humankind. It is known by the contranym 'synthetic biology'.

Definitions in science are often imprecise, and frequently unhelpful. Synthetic biology is a term that means different things to different people, and aspects of those definitions are discussed in the following pages. Scientifically, it is a direct descendant of genetic engineering, and the overlap between the two is not always clear. For that reason, I have looked at both in the course of exploring our newfound ability to create life forms using parts of genes and cells taken from the toolbox that evolution has provided. And finally, as a postscript, I will explore the creation of entirely new molecules that fit into DNA but are not part of evolution's lexicon – new languages and new uses for the genetic code.

All of the creations of genetic engineering and synthetic biology are new to the inventory of life. Most of them feature minor

modifications, or small additions that add a new function to (or indeed knock an existing function out of) an organism. Almost all occur in bacteria, whose genetics are understood well enough that they have become essential equipment of the modern era of biology. Even so, synthetic biology is a new field, and many people may not have come across it before.

That changed in May 2010, when the news was briefly – and unprecedentedly – besieged by a single cell. It was given a name, and photos of Synthia, as it became known, adorned front pages and television channels all over the world. After more than ten years of toil, and roughly $40 million, the geneticist J. Craig Venter published a paper that described a bacterial cell that his team had created. Its genome had been assembled not inside a parent cell, as with every other cell in history, but in a computer. The press swooned. One magazine put Venter as the fourteenth most influential person on Earth, sandwiched between the current British prime minister David Cameron and the American politician Sarah Palin. Although it was a major landmark, Synthia was not quite part of the synthetic biology revolution that is now beginning.

Craig Venter himself was keen to mark this event in grand terms. To illustrate our command over DNA, and to distinguish Synthia – aka *Mycoplasma micoides JCVI-syn1.0* – from naturally occurring bacteria, Venter and his team hid what video game players call an 'Easter egg' in its genome. This was in the form of secret encrypted messages in the DNA of his creation. One of these was 'What I cannot build, I cannot understand,' which fractionally misquotes Feynman. An unnamed member of Craig Venter's team had obtained it from an unknown site on the internet, a reference source which, on occasion, has been known to be wrong. But, errors aside, this Easter egg quotation served the purpose of stamping an indelible 'watermark' into this bacteria, proving that it was unnatural in origin. It also showed that DNA can act as a data storage device, a means of encoding information that is not biologically relevant. We shall come to this emerging field in the Afterword. Overall, this whole story was an impressive demonstration of just

how far we have come in our technological ability to manipulate and build DNA.★

Synthia owes its existence to the Minimal Genome Project. *Mycoplasma genitalium* is a parasite that causes minor burning and itching in infected men and women when they pee. That's not what is interesting about it, though. Venter's interest in this cell is because it has a mere 517 genes, and a genome of only 582,000 letters of genetic code, called bases (compared with 4.6 million bases for the more common lab microbe *E. coli*, or three billion for humans). After sequencing the first complete genome in 1995, Venter and his colleagues' second quarry was *M. genitalium*. This choice was no act of laziness, but the critical first step in Venter's next big idea. The plan was (and is) to determine the smallest number of genes and the minimum amount of genetic code required to sustain a living cell. Life originated with only a handful of basic genes, and they were copied and mutated for all subsequent life to flourish. But those first cells must have started with an entry-level set. Starting with the smallest known genome (at the time) made a lot of sense. The long-term plan was to establish the minimal amount of DNA required for a cell to exist and reproduce, so that they could use that genome as a foundation on to which they could build new functions. Those functions are typically grand: for example, to address the environmental damage that our use of fossil fuels has brought upon the planet by building microbes that

---

★ Synthia contained three hidden quotations, about which British satirist Charlie Brooker commented: 'the geneticists spliced a James Joyce quotation into the DNA sequence. The unsuspecting genome now has the phrase "to live, to err, to fall, to triumph, to recreate life out of life" written through it like letters in a stick of rock. In other words, it's the world's most pretentious bacterium.' That quotation was lifted from the novel *Portrait of the Artist as a Young Man*, but Venter and his team neglected to seek permission from James Joyce's estate, who are notorious for aggressively protecting their copyright. Executors of Joyce's estate immediately issued Venter with a cease and desist letter. In fact, the copyright had expired, and no further action was taken.

might produce hydrogen for fuel. There are a few reasons for building a cell from scratch rather than modifying an existing one. They can be controlled more precisely in defined conditions, and therefore only grown in labs; they can have singular purpose built into their genetic program so that the normal functions of cells are curtailed. But the primary reason is control: we can potentially control a synthetic cell far better than natural cells that we don't fully understand.

Synthia represents a proof of principle, that it is possible to construct an entire genome synthetically, albeit a very small one. And it is possible to get it into a cell, so that the cell functions and, crucially, reproduces like any other bacteria. More broadly, this saga shows the ambition of scientists to turn biology into an applied science, into engineering. It indicates that the molecular manufacturing required in synthetic biology is not at all easy, which is not surprising, given how young this field is. The protagonists of the purest form of synthetic biology have engineering as their key guiding principle and, even more specifically, the commoditized version of electrical engineering. These aims are not merely to investigate and understand how living processes function, but to re-create, remix and build living organisms that address global problems.

With the publication of Venter's creation, the media brandished a curious mixture of teeth gnashing and misplaced triumphalism. The *Daily Mail*, a UK newspaper not known for nuanced reporting on scientific advances, ran the story with a headline that bellowed: 'Scientist accused of playing God after creating artificial life by making designer microbe from scratch – but could it wipe out humanity?' 'No' is the very straightforward answer to that question, as is inevitably the case when newspaper science headlines end with a question mark. Here is why. Synthia's genome was designed with failsafe devices built into its genome. In general, these can work in two ways: you can include genes which mean that your modified bacteria can only grow in a very specific

blend of food that only the lab knows how to produce. Second, the genome from which Synthia's was copied and modified is based on a minor goat pathogen, one that causes udder infections. Venter's team inserted a chunk of DNA that renders the cell incapable of causing infection. So while it might be able to just about survive outside of the lab in which it was created, it would not be able to thrive in its preferred habitat unless someone with molecular biology expertise further modified its genetic limitations. Could it wipe out humanity? No, but if a skilled geneticist was persistent and mean-spirited enough, it might well bother a goat.

The brouhaha was not limited to tabloid newspapers. Eminent professors weren't shy in joining in the hubris. Julian Savulescu, professor of ethics at Oxford University, told the *Guardian* newspaper:

> Venter is creaking open the most profound door in humanity's history, potentially peeking into its destiny. He is not merely copying life artificially . . . or modifying it radically by genetic engineering. He is going towards the role of a god: creating artificial life that could never have existed naturally.

Of course, a Friesian cow or a Savoy cabbage are both life forms that are extremely unlikely to have existed naturally. They have been bred over innumerable generations for the express purpose of amplifying desirable characteristics: milky udders or a crunchy taste. Though I doubt anyone would accuse your average farmer of creaking at profound doors.

Did Venter create life? In a very particular sense, yes. But in most others, no. Certainly Synthia did not exist as a living thing until Venter's team booted it up in a plastic tube. Its genome was created on a computer and then in a machine fed with four bottles containing the raw letters of genetic code. It was not derived from the biological replication of an existing genome copying itself using the method that has served life for billions of years. But the sequence itself, modified with genetic failsafe suicide switches and

pretentious (mis)quotations, was replicated from an existing species. And the chassis into which the synthetic genome was placed was from an existing cell from a naturally existing species, one distinctly embedded in the firmly rooted tree of life. What Craig Venter and his team did was to re-create a life form synthetically. That is undoubtedly a huge achievement in itself. It's another incremental step on the pathway to having total control over DNA, and our ability to manipulate life.

The second thing that Venter's endeavour reveals is quite how poorly understood and misrepresented this nascent field is. We have been manipulating living things for tens of thousands of years, though never before with such molecular precision. The pace of progress has been breathtaking. A decade ago, I finished my Ph.D. in genetics, during which time I was doing minor acts of genetic tinkering in order to work out how the eye develops, and why in some families it goes wrong, leading to children born blind. The types of experiments I was doing then can now be done by undergraduate scientists and even hobbyists known as 'biohackers' – amateurs, sometimes of school age, who mix the genes and DNA of organisms in garages on the weekend. This in one sense is natural progress, where new technologies become normalized and the exclusivity that comes with expertise or invention is lost as they become standardized tools. But the pace of that transfer is daunting, and poorly understood outside of the world of science. And in the right (or worse, wrong) atmosphere, people fear what they don't understand. As the story of Craig Venter and Synthia shows, extraordinary science can result in visceral reactions.

What follows is a snapshot of this embryonic field of the engineering of living things. The synthetic biologists have application in mind, engineered and constructed solutions to the problems that this planet faces and, more incredibly, the tools to explore beyond our earthly shackles. Many of the applications of the creations described here are speculative, but they are nevertheless built

with purpose in mind. With the advent of these creations, we stand on the precipice of a new industrial revolution. The wood, iron and coal that drove the industrial revolution of the eighteenth and nineteenth centuries were indeed drawn from nature – cut, mined and extracted from the world. But in use they were rendered inanimate and dead. This time round, the agents of change are not just taken from the natural world – they are alive.

1.   Created, Not Begotten

'I am against nature. I don't dig nature at all.
I think nature is very unnatural.'

Bob Dylan

Freckles looks like a perfectly normal kid. She has bright eyes, a healthy white pelt, and gambols happily with Pudding, Sweetie and her five other siblings, exactly as you might imagine a young goat would. Until I fend her off, she's very keen on chewing on my trousers. To the casual observer, and to professional goatherds, she shows no signs that she is not a perfectly normal farmyard goat.

In fact, Freckles is utterly extraordinary. While almost all of her genome is what you would expect a goat to have, there is a foreign corner in her DNA taken from *Nephila clavipes*, the golden orb weaver spider. The language of the DNA is exactly the same in both goats and spiders, as it is in all living things, but the translation of that particular piece of code is very alien to your average goat. It has been inserted at a very specific point in Freckles' genome, adjacent to a coded instruction that prompts the production of milk in her udders. As a result of this genetic intrusion, when Freckles lactates, her milk is replete with spider silk.

Freckles is the creation of a team of scientists from Utah State University led by Randy Lewis. I visited their lab, which is in fact a farm at the base of a daunting mountain range in Logan, Utah. Though a long way from the sterile clean-rooms of a typical molecular biology lab, this setting is apt enough, as Lewis and his scientist colleagues are in one sense the most advanced farmers in the history of our species. It's dragline silk that they are interested

in, the gossamer fibre that spiders use to swing from when falling. This material has stunning physical properties, with a combination of strength and elasticity unsurpassed by anything that humans can yet manufacture. The orb weaver genus has existed since the Jurassic period, which began around 201 million years ago, and during the epochs before and since evolution has conspired to bestow a product on these spiders that is so far beyond human-kind's greatest efforts, but which the spider produces in bulk without a thought. We can make fibres with strength or with great elasticity, but not with both. That is why Freckles exists, the unnatural creation of scientists.

For more than 10,000 years, since the dawn of agriculture, we have identified appealing characteristics of the natural world, and attempted to enhance and exploit those properties. This is farm-ing. It is a process that is at heart the precise opposite of natural selection, though it harnesses the exact same means. Instead of survival determining what unintelligently designed adaptations evolve, we choose characteristics that we find attractive and breed species to optimize production of that trait, whether it is fruit or arable, or livestock for meat, dairy or leather, or dogs and plants bred for behaviour or aesthetics. Farming is evolution by design.

Until the era of genetic engineering, this artificial selection has been inherently limited by natural boundaries. Although notori-ously difficult to pin down precisely, one definition of a species is a group of organisms that when crossed with each other can pro-duce fertile offspring. It's not a wholly infallible definition, and the process by which that barrier is introduced – speciation – is one of the great questions in biology. But from the point of view of a farmer, the limitations of the species barrier are unsurpassable. In crude terms, you can't mate a pig with a cow.

Desirable traits are introduced over time by careful selective mating over generations, a fact well known to Charles Darwin. The opening chapter of *On the Origin of Species* is devoted not to his central thesis of evolution by natural selection, but to artificial selection in pigeons. In this pigeon section Darwin is demonstrating

that species are not immutable, that they can change over generational time. The birds he describes had been reared over thousands of generations by competitive fancy breeders to feature ever more preposterous prize-winning plumes, gullets and feet, such that they had names like Trumpeters, Fantails, Tumblers and Pouters. Through meticulous observation of their skeletons, and in direct opposition to the forcefully held opinions of the pigeon men, Darwin correctly revealed that these pigeons were all still varieties of the same species, the rock dove *Columba livia*.

With the discovery of the genetic code – DNA – and with the advent of our ability to manipulate it, we are now capable of fully circumventing the limitations of evolution and radically violating the species barrier. Freckles, though by no means extraordinary in terms of genetic modification techniques available now, is a dramatic example of this biological infringement. Pigs and cows are both mammals, both even-toed ungulates and closely related in evolutionary terms to each other and to camels, giraffes and hippopotamuses. Their common ancestors existed a few tens of million years ago, an evolutionary blink of the eye. But that distance is enough to forefend viable offspring if they were to attempt a mating. This barrier is laid down by diverging genetic incompatibilities, though the physical congress is conceivable. A few closely related beasts do produce viable children, such as zeedonks, ligers and hinnies, though these hybrids tend to be infertile and therefore evolutionary cul de sacs as they cannot themselves reproduce.*

---

*Zeedonks (aka zebonkeys, zebrinnies, zebrulas and zebadonks) are a cross between a zebra and a donkey. Similarly, a liger is a lion and tiger cross, and a hinny a male horse and a female donkey, as opposed to a mule, which is the opposite. Sterility in these unusual hybrids is in accordance with a principle called Haldane's Rule (after J. B. S. Haldane, who came up with the term 'primordial soup'), which states that, when a cross is successful between two species, one sex will be absent in the offspring, and that will be the one with two different sex chromosomes, that is XY (males) instead of XX (females), in mammals. In birds and butterflies, this is reversed.

The last common ancestor of a spider and a goat would have existed something like 700 million years ago, at a time when the things that would acquire hard shells, such as insects and crustaceans, were evolving away from creatures with fleshy exteriors, such as the fishy or reptilian beasts that would eventually lead to us. As the tree of complex life only grows via diverging branches, spiders and goats have not exchanged genes or DNA since. Unlike a zebra and a donkey (or even a pig and a cow for that matter), sexual union of a spider and goat is obviously physically impossible. But the DNA code they both carry is the same across all known life forms. It is the same language, and formatted in the same way, so that the tools and mechanics of the translation of that code are blind to its cryptic meaning. So, if a spider dragline silk gene is introduced into a goat genome in a way careful enough not to disrupt essential biology, the goat's cellular machinery will produce spider silk without reference to its aberrant origin.

And that is exactly what happens in Freckles' udders. In spiders, dragline silk is made of short strands of protein, which align and self-assemble as they are pushed or drawn out of the spider's spinneret. Proteins themselves are long strings of molecules called amino acids, and the very specific combinations of amino acids in silk fibres mean that they line up and overlap, locking into continuous threads that have built-in give. Because of this, they have length, strength and elasticity. Randy Lewis' goats produce the short silk proteins in abundance, which float freely in the milk. The fat is removed from the milk, which is then pushed through a high-pressure processor. From there, simply with a glass rod, I lift a single thread of spider silk from the goat's milk, and the noses and tails of the shorter fibres link together as I draw them from the liquid. It is strong enough to wind onto a spool by the metre. As yet, this unnatural silk does not quite have the same elastic or tensile strength as that created by spiders themselves. The research continues.

Although seemingly bizarre, this process is effectively next-generation farming. Its physical properties make spider silk an

extremely desirable substance for humankind. It is tougher than the man-made fibre Kevlar that is used in body armour and puncture-proof tyres. It is also a substance that does not generate a significant immune system response, and it is insoluble in water, two facts that the classical Greeks exploited when they used cobwebs to patch bleeding wounds. These properties mean that spider silk has great potential for ligament repair, for which medicine currently uses slivers of flesh extracted from the patient's own muscles, or tissue from dead bodies. Neither of these solutions is permanent, so we could do with a replacement ligament made of a biologically neutral substance that has physical strength equal to or better than our own ligaments.

It might seem absurd to say it, but spiders cannot be farmed. When kept in groups they have a tendency towards cannibalism. With our ability to violate natural species boundaries using the tools of modern genetics, we are able to harvest a product by placing its code in a farmable animal. Although Freckles the spider-goat is a striking example of our prowess with gene mixing, she is not at all unique. This is genetic engineering: the creation of unnatural life using evolution's bountiful toolbox.

The language of DNA is carried by all living things because life on Earth ultimately evolved from a single origin, a cell that existed around four billion years ago.* The language, code and mechanics of all living cells are the same. DNA bears a hidden language, which is decoded via a messenger molecule called RNA, which is translated into proteins. Proteins carry out all the functions of life, as enzymes that speed up essential biochemical reactions inside our cells to provide us with vital energy, or as missives such as hormones that carry instructions from one part of the body to another. Proteins form structures that hold our cells together, and they fabricate other structures such as bone, teeth and hair. In short, all life is built from or by proteins.

---

* This cell is known as Luca – the Last Universal Common Ancestor, and is discussed at length in 'The Origin of Life'.

Because of this common origin and universal process, the cellular mechanics of reading and translating the code of DNA are both blind and indifferent to its meaning, which means that the code of living things can be used to transgress the natural barriers of reproduction. Chimeras such as Freckles are called 'transgenic' organisms, as the genes have been moved across the species barrier.

Scientific progress always relies on the development of new technologies. Humans excel at tool use, and we have always looked to nature for useful instruments. We see what is available to us, and appropriate and modify it for different uses, all the way back to using bones as clubs. The natural world has always served as our toolbox. We chipped flint into crude knives, carved tree branches into arrowheads, smelted metals from rocks, and wore animal skins sewn with bone needles and plant-twine. The technology that enabled the violation of the species barrier that genetic modification requires is also borrowed from nature's staggering bank of resources. They come from our most distant evolutionary cousins: bacteria.

In 1968 Hamilton Smith (then working in the Microbiology Department at Johns Hopkins University in Baltimore) isolated the first of a set of proteins that launched the era of genetic engineering, and for this he shared the 1978 Nobel Prize for Physiology or Medicine. These proteins are known as 'restriction enzymes' and very simply they act like DNA scissors. In nature, a bacteria's restriction enzymes cut their own double helix as a means of protecting against the invasion of viruses. As a virus carries only its genetic code, and none of the machinery to replicate it, its purpose is to hijack the host cell to reproduce. The infected cell translates the virus' code and unwittingly makes new viruses, releasing them en masse until the cell bursts. Restriction enzymes are part of the arsenal to prevent that hijack, by editing out the interloping DNA.

The ability to cut DNA is of profound use to geneticists, as restriction enzymes do not snip the code randomly. Instead they

only cut when they recognize a precise row of letters in the genetic code. There is a huge range of restriction enzymes, with some cutting at common sequences, and some at very rare pieces of DNA. Imagine that working in this book: a restriction enzyme that makes a cut when it sees the word 'cell' would shred this book into fragments, as if blasted with a shotgun. A restriction enzyme that cut when it recognized the word 'jumentous' would chop the whole text neatly into just three pieces, as that uncommon word appears only twice.

This naturally occurring editing tool enabled scientists to treat DNA and genes as we now use word processing software: to cut, copy and paste from one section to another. This paragraph began its life in another chapter, but with a couple of simple clicks, I transposed it here, and modified it to make sense. Similarly, the dragline silk gene began its life in a spider, and was transferred into the genome of a nascent goat, tweaked and positioned so that it makes biological sense. Hundreds of restriction enzymes have now been isolated and characterized, and are cheap and readily available. Even better than merely moving text around, as DNA is a double helix, made of two strands that mirror each other with complementary code, there is another useful trick that these genome-editing tools can perform. Some simply cleave both strands of the DNA at the same point. These are called 'blunt' ends, and any other blunt-ended DNA can readily stick to it, two ends glued together, like a blank domino. They are perfectly stable once linked, and have value to the gene engineer as you can stick them in anywhere. But other restriction enzymes cut each strand a few letters apart, creating 'sticky' ends, with one strand of DNA overhanging the other. That makes them an uneven domino, where you can only join to the end if you have the matching piece. A stretch of alien DNA that has the complementary sticky end can join up into the host DNA, clicking neatly into place. This means inserts can be orientated in a particular direction, and other useful inserts added to make more complex genetic pathways, a run of dominos built by design.

In the lab, almost all of these molecular biology experiments involve transferring minute quantities of colourless liquids from one tiny tube to another. The action of restriction enzymes on DNA is invisible to the naked eye, and we check success with a battery of proxy visualizations, such as measuring the mass of the cleaved or rejoined DNA fragments. It's often far from dramatic, and not easy to see the profoundly unnatural acts being perpetrated. Ultimately, though, the real test is to see whether and how these reconstructed codes work in cells, and in organisms. Sometimes the effects might be gross and visible. But frequently the impact of genetic modification might be subtle or cryptic, and so being able to detect where and when your experiment is working is a fine art.

In genetics, making your transgene, the code with which you are messing, is only the first step. The code does nothing on its own, and so it has to be placed into a cell where decoding will occur, and you hope the subverted message will be enacted. In Freckles the code is not merely the spider silk gene, but instructions that say when it will be active, or 'expressed' in the language of genetics. By careful construction, the silk will only be produced in the milk, even though the gene is present in every one of her cells.

During the 1990s, extra bits of genetic code added to the toolshed included coloured tags so that down the microscope we could see exactly whether the gene had been successfully taken up into its new carrier. More recently, fluorescent tags have been added so that we can see where the gene is active once it has been integrated into an animal. The most visually striking example of this is the functionally named Green Fluorescent Protein (GFP), which in its natural host, the jellyfish *Aequorea*, glows green in the pitch-dark oceans. The gene that encodes this beacon was extracted from jellyfish, so that it can be added onto the gene you are testing. Just as in the spider-goats, the machinery that translates the DNA into working proteins is indifferent to its meaning, so reads the

universal genetic code without species prejudice, and glows to show that your modified gene is working.★

Genes never work in solitude. Instead they are parts in cascades of networked activity, like a series of conditional instructions. Every cell contains every gene in that organism's genome, regardless of whether it is needed. So the choreography of gene activation is of paramount importance. The activation of one gene might trigger the activation – or expression – of another, or tell it to stop, and by this process we develop from a single fertilized egg cell to a collection of hundreds of coordinated cell types, each performing different functions as a result of the expression of specific sets of genes, rather than an amorphous blob comprising identical cells. This dance is elegantly choreographed, but with limited flexibility. A gene activated at the wrong time, or in the wrong place, or for too long, can lead to disease, abnormality or often fatality. Cancer cells are ones in which genes that would normally tell other genes to stop cell division have failed (or conversely, genes that tell the cell to reproduce are permanently active), so the cell only divides. This uncontrolled reproduction grows into a tumour.

The command to activate a specific gene will typically come in the form of a protein physically attaching itself not directly to the gene itself but to the DNA nearby. These instructions are called regulatory regions, like instructions for flat-pack furniture. So one gene will produce a protein that will bind to a particular run of

---

★There is plenty of artisan skill required to get the additional DNA to function. The syntax and phrasing of the inserted gene has to make sense to the cell's translation machinery. The arrangement of the letters of DNA – A, T, C and G – is precise and crucial to the protein that a gene encodes. Those letters, or more properly bases, are set in threes, each triplet encoding an amino acid, a kind of biological waltz. If your insert comes in on the wrong beat, the dance makes no sense. But when you introduce the jellyfish gene so it starts on the right beat, wherever your experimental gene is active that cell will glow green. This tag is small enough that it does not affect the function of the protein, just like someone wearing a brightly coloured hat in the waltz.

DNA bases near another gene telling it to become active, and a continuous network of activity buzzes inside each cell and between cells to create a life. A large amount of most genomes is not genes, that is, DNA that encodes a protein. Genomes are littered with these regulatory regions, which don't contain code themselves, but act instead as very precise stage directions for the proteins to enact their role.

Through the techniques of DNA modification, turning these networks of instructions into tools has rendered molecular biology an experimental science. And, as is so often the case in biology, we can determine how things work by breaking them. Traditionally this has taken the form of studying hereditary diseases or by looking at mutated animals. As mice and men share many genes, we can knock out a gene in a mouse embryo by modifying its DNA to see what the result is. This is done using restriction enzymes and imported regulatory bits of DNA and spliced-in transgenes, together forming a process of major reconstructive DNA editing.

Extraordinary though they may sound, these techniques have been the meat and drink of genetics for more than a decade, and the industrialization of the process of gene splicing and editing has made genetic experimentation easier. We have taken the tools provided by evolution and modified, designed, redesigned and honed them to provide service beyond the roles they play as a result of natural selection. A typical experiment to determine the function of a human disease-causing gene would be to isolate it in a family that suffers from that disease, using the same principles of genetics and pedigrees that have been known for a century. Once purified, it gets copied a million times, so there is more DNA to play with. That makes inserting the modified gene into a precise orientation in a DNA envelope much easier, and that can then be posted into a bacteria. A tiny burst of electricity will punch temporary holes in the bacteria's cell membrane, and the constructed DNA will just flow in, and with skilful design and a bit of luck it will be incorporated into the host's genome. Every time the cell then

divides, the inserted gene will be copied along with the host genome. This is a numbers game, so bacteria are the perfect tool not
just because we can manipulate them but because they grow like
crazy. The next step is to destroy the bacteria and leave only your
modified DNA. At that point, you can do what you want with it.
You can make RNA versions of it, which will show where the
gene is active in preserved tissue on a slide, in an organ or even in
a whole animal. Or you can put that gene and all its extra controls
into a mouse embryonic stem cell, and implant that into a mother
to see what it does as the mouse develops.*

None of that is exceptional, just standard molecular biology
going on in thousands of labs all over the world. We have become
so adept at using the tools and the language at the core of all living
things that manipulating living systems at their most fundamental
level is trivially easy. Through experimentation, we know how
genes in the cells work and, more importantly, how we can change
them. We can correct genes that cause diseases, not yet in humans
but in animals, and that allows us to study how diseases progress,
and therefore how we can treat them. We have developed the ability to read and characterize every single gene in a human, at the
time of writing, in a few weeks at a cost of a few thousand pounds
(though both numbers will inevitably continue to fall).

Synthetic biology is the evolved descendant of genetic

---

* This process is called 'cloning' but is unrelated to the popular sense of that
word, meaning to reproduce an identical organism as in science fiction or in
animals such as Dolly the Sheep. My own, atomically minuscule contribution
to the field of genetics was in a team that worked on a mouse born naturally
blind, with shrunken eyes, a deformed retina and no optic nerve. By various
techniques, our team, led by Jane Sowden at the Institute of Child Health in
London, excised and spliced various genes and genetic elements to characterize
what the faulty gene was doing and when. Using the basics of cloning, we cut
the gene out of the genome of the mouse and used it to determine the precise
time and place where it should have been working. We tagged it and looked for
the other genes it was controlling and being controlled by. By doing this, we
helped to discover that the equivalent gene in humans also causes a rare form of
blindness in children.

engineering. It takes the principles of biology and reinvents them with the goal of engineering solutions to specific human problems – disease, environmental issues and, as we shall see, even space exploration. This synthetic biology movement is a new phenomenon in science, a decade old at the most generous estimate. It has carried with it a subculture ethos too. But, unsurprisingly, mainstream science and corporations have also begun to take notice of its potential.

Climate change and global warming will define much of synthetic biology's industry and innovation in the next few decades. Expecting people to radically change their behaviour is unrealistic, so finding alternatives to fossil fuels is a major challenge. But in the meantime synthesizing fuels, rather than digging them out of the ground, is a partial solution.

There are dozens of projects to make biofuels, converting vegetation into energy in that fundamental way that so many life forms do naturally.* It is a process of carbon fixation, effectively turning carbon dioxide into organic, energy-rich products. In nature, the energy of the sun is harnessed in various metabolic processes by a plant so that it can live and grow. Setting fire to a crop will release that energy, now stored in the plants' strong cellulose cell walls. But a more efficient way of tapping into that energy is to ferment the sugars locked in plant cells, also the product of the sun's energy, directly into an oil fuel, which is crammed full of more readily available energy. A handful of companies are attempting to make biofuel from synthetic biology using this very process.

One such project is run by Jay Keasling, a professor at the University of California, Berkeley, who founded Amyris, a synthetic biology company whose aim is to create diesel from live cells. Their designed tool has been a genetic circuit consisting of around a dozen individual chunks of DNA, which they implant into the

---

* Petrol itself is a kind of biofuel: crude oil is made of the carcasses of trillions of organisms which have decomposed and been crushed for hundreds of millions of years.

genome of brewer's yeast, a cell that naturally ferments sugar into alcohol in beer. The success of this circuit in producing diesel is no mean feat, and the scale of their ambition is mesmerizing.

I have been in scores of genetics labs, which range from the charmingly old-fashioned to modern office space. Molecular biology doesn't necessarily need the huge clinical white spaces that the movies portray. They are essentially very precise kitchens, with ovens and fridges and tools for mixing ingredients. But Amyris' labs are stunning. Only the very deep pockets of a commercial venture can provide such luxury, not merely in its space and light, and a healthy San Franciscan café, but in developing the kind of technology that is required in order to industrialize synthetic biology.

One of the most time-consuming tasks in a normal lab is the insertion of the genetic circuit into the host cells. The efficiency of this process is variable. It's not random, but there is an element of hit or miss: the circuit either goes in or it doesn't. In normal lab gene tinkering, we include genetic markers in the circuits, tags that colour the cells to indicate whether integration of the alien DNA has been successful. In order to see those colours, the cells have to be grown not in a broth but on a plate, so that the successful colonies of cells can be selected with a toothpick and grown in a small tube, leaving the failures for the incinerator. It's a trivial but arduous process and it is done by hand. At Amyris, they have mechanized it so that they can pick hundreds of successful clones in minutes, tens of thousands per week, with only minimal human input. A digital camera takes a snap of the plate that bears the thousands of yeast colonies and processes it to identify the successful clones. Inside a machine as big as an office photocopier, a bank of needles hovers and zooms over the plate, plucking the successful yeast cells with sub-millimetre accuracy at the rate of a disco beat.

The cells that have taken up the circuit don't need much more encouragement. Incubated, they simply leak diesel. At Amyris' headquarters in San Francisco they have a 300-litre test tank churning out diesel by the litre. The labs smell sickly sweet, like a

brewery, but more appley, which is because the diesel they are refining, called farnesene, is found in the oil that gives apples their water-repellent skin. As a fuel, it is cleaner than petrol-based diesel as it has no sulphur emissions and reduced nitrous oxides and carbon monoxide.

Creating diesel affords us certain advantages over digging it out of the ground. But it also comes with a different problem. Set aside the fact that clean, synthetic diesel is still diesel, so it is still a carbon-dioxide-generating fuel. The more significant problem is that the vats of yeast seeping out diesel need to be fed. That means converting plant material, biomass, into feed. Growing that biomass is a problem of traditional farming, and so comes with the same issues that farming does. Amyris have made strong partnerships with companies in Brazil, to be close to where the first-choice food is grown, which is sugar cane. Brazil is the ideal location for this project as they are the biggest farmers of sugar cane, and since the 1970s have produced biofuels, predominantly ethanol, as a means of freeing themselves from dependence on foreign oil. The key question is this: how much land does it take to make one litre of diesel? The answer is unclear at this point. Different feedstock is digested in different ways, and the synthetic yeast programmes can be altered to cope with this. In their Brazilian plants, synthetic farnesene diesel is already being supplied to local vehicles and as an alternative aviation fuel. But Amyris' goal was to grow output to 200 million litres by 2011 at $2 a gallon. The ambition is clear: Jack Newman, Amyris co-founder and chief scientist, told me, 'I'll be excited by a billion litres.'

But it looks like Newman's billion litres is for now a burst bubble. For a while, Amyris looked like they were going to be the winners in the race to make and sell commercially viable synthetic biofuels. But then the 200 million litre prediction dropped to fifty million for 2012. And in February 2012 they announced that they were scaling back production of farnesene until the industrialization is more plausible. The genetic program works just fine, but scaling it up to the levels where it becomes economically viable so

far does not. For now, for all the stunning laboratories, it seems that the synthetic biofuelled future is still a long way down the road.

Meanwhile, we have been able to modify crops to be resistant to blights, grow larger, tolerate frost and even produce vitamins that protect consumers from diseases. By borrowing genes from a bacteria and a daffodil, scientists have created a form of rice that produces high levels of beta-carotene, a molecule involved in Vitamin A production. Golden Rice, as it is called, has the potential to treat the 120 million people worldwide who suffer from Vitamin A deficiency, two million of whom die and half a million of whom go blind. But it remains unavailable due to a combination of scientific and ethical roadblocks and politics.

These stories show some of the promise, potential and problems with synthetic biology. In essence, it is engineering, with application of basic research at its core. But, alongside its forefather genetic engineering, these are industries that are immature. They face scientific problems, commercialization problems, upscaling problems, ethical problems and, as we shall see, dogged resistance from some members of society. Biology is messy, and complex genetic networks underlie that mess. As genetic engineering, only three decades old, gives way to synthetic biology, the great challenge is not merely to simplify those networks, but to commoditize them.

2.   Logic in Life

'if it was so, it might be; and if it were so, it would be;
but as it isn't, it ain't. That's logic.'

Tweedledee, in *Through the Looking Glass* by Lewis Carroll

Flick the switch, the light comes on. This is the simplest useful electrical circuit. The parts are designed and created to follow one single instruction: the switch is a gap when open, but when closed electrical energy surges through the circuit. The filament in the bulb converts some of the electrical energy into a form that we can detect with the cells in our eyes, and so enlightenment occurs. The function is clear, the instruction pure, the logic is unimpeachable, the light comes on.

At the other end of the scale, you watch a video streamed over the Internet on a laptop computer. Billions of electrical signals will have been created, modified and transmitted in order for that moving image to play. The circuits have been designed in intricate detail with each one obeying a logical pattern determined by the hardware and software in your mouse, your computer, the servers that host the file, and so on. The logic is also perfectly clear, but the complexity of the pathway makes it all but inscrutable to almost everyone, and occasionally unpredictable. And yet we use the output of this tangled circuit every day without care or understanding of the millions of decisions that have been made to put a moving image on your screen.

Recall, if you can bear to, some of the electrical circuitry you learned at school. A light bulb connected to a battery with a simple switch has a binary output: ON or OFF. You may have then

added other devices into the circuit to introduce finer levels of control, such as diodes – electrical valves that turn a wire into a one-way street. Add, for example, a thyristor, and you have a dimmer switch. Possibly the greatest invention, certainly the most enabling technology, of the twentieth century was the transistor, which allowed the ability to control and modify multiple electrical signals. Interconnected transistors make up what are known as logic gates, which modify the input signal to produce a specific but different output. With the introduction of logic gates, circuits of ever-increasing complexity can be designed and built. For example, an AND gate acts as a positive conjunction: if two electrical inputs feed into an AND gate both must be ON in order for the output to also be ON (in electronic engineering, capital letters indicate logical calculations, rather than merely apparently bellowing for emphasis). A microwave oven uses this logic. It will only cook if the door is closed and the start button has been pressed. If either of the two crucial signals are negative, then the output is negative.

Electrical circuitry relies on logic, on following pathways determined by the component parts. From the light switch all the way up to the machine I am typing on, the route of information takes the form of digital questions with digital answers: if I press the on switch the light comes on; if I press return (via a long pathway of transistors and thousands of other components) the paragraph ends.

These are the very basics of electrical engineering, a field that has grown from light bulbs to relaying messages to and from deep space in little over a hundred years. Every signal sent to your mobile phone, or from the spacecraft Voyager 1,* is relayed via the logic ingrained in transistors. In a standard microchip, there will

---

* Voyager 1 is the most distant man-made object from Earth, and at the time of writing is eighteen billion kilometres from us. It sends regular messages via Twitter as @nasavoyager.

be billions of transistors, and this is the technology on which almost everyone now relies.

It's a hugely appealing system, not just because it works as we design it to, but because the fact that it works has allowed us to build the modern world. This logic and this ambition is at the heart of synthetic biology. Many components have been built, and some have been assembled into basic circuitry. Although the term 'synthetic biology' has gained broad usage to include everything from the work of Craig Venter on Synthia to the reinvention of genetic code described later, the field originated and has been maintained by scientists aiming to apply the principles of engineering, specifically electrical engineering, to biology. Living things are immensely complex, with thousands of genes encoding thousands more proteins which interact with each other and the environment to produce millions of cells. But the basic logic of genetics is present in principle: if a gene is activated, the protein it encodes will be activated, and will perform a function.

We think of life forms not as illogical, but not with the straightforward formulas of electronics. Nevertheless, the principles of logic gates as they are used in electronics do exist in nature, as complex behaviours and actions often need the parsing of multiple inputs. The carnivorous plant the Venus flytrap displays a simple but clever form of logic circuitry when performing its eponymous act. On the inside of the leafy jaws are tiny hairs that act as the triggers for clamping shut. But that act requires metabolic energy, so it has evolved a mechanism for avoiding fruitless (or flyless) trapping. When a hair is touched by a fly's nudge, a timer begins. If a second hair is triggered within twenty seconds, the jaws snap shut in less than one-tenth of a second. Thus the input signals that result in the action of trapping are required in conjunction. The double trigger is a form of AND gate, as both outputs from the hair triggers need to be set to 'ON' for the output of the full circuit to be 'ON'. There's also the timer built into each trigger, so the overall simplified electrical pathway might read: IF TRIGGER 1 + TRIGGER 2 < 20" THEN CLOSE JAWS.

This process is, of course, managed and run by the mechanics inside the Venus fly trap's cells, with specific proteins that respond to the physical sensation of a fly's touch. But the biological action of a trap waiting to be sprung follows a straightforward logic that looks a lot like electronics: it has no brain or conscious control for the decisions it makes, it just follows a program.*

But mostly living things follow a logic that is a far more difficult lock to unpick. The activation of genes to produce functional proteins is bound by time and location. Genes interact and respond to the environment in which they are active, in the cell, and from long- and short-range signals from close and distant neighbours. Genes themselves vary from person to person and from creature to creature in small and subtle ways. This can make their function, their output, also vary from person to person. This variation is essential as it is the basis on which natural selection can act so that evolution unfolds. Combine that with the muddled interplay of genes and environment and we have absolute individuality. This is the reason why fingerprints are unique, even on the fingers of identical twins, whose genes begin their existence identical.

There are some cells that work in a basically digital way. Brain cells, neurons, fire with a dynamic flourish, and link up to produce thought and sense. Their spark is only initiated when inputs from other cells (in the form of charged atoms) flow into the neuron until a particular threshold is reached. This process, with a suitably dramatic name – action potential – renders these neurons digital: they are OFF until they are ON. But don't let that simplicity fool you into thinking your brain is easily understandable. There are more than a hundred billion cells in your brain, and each one of them forms thousands of connections with others. This means potentially hundreds of trillions of switches flicking on and off in

---

* The Venus fly trap circuit is qualitatively different from electronics because the logic of the genes underlying the process is there not to enact the trapping, but to set the trap, that is, to open the jaws and arm the trigger. They express the proteins that act as the components in the logic pathway. In electronics the logic is dictated by the flow of electricity through the circuit.

millisecond bursts every moment you are alive. There may be a logic buried in there, but for the foreseeable future it is inscrutably complex.

For the most part, the switches in life are not quite so binary either. There are hundreds of other types of cells and switches that are much more nuanced, stimulated on a sliding gauge, or contingent on myriad inputs. Just like a neuron, a gene may indeed be 'ON' or 'OFF' like a light bulb. But it also may be 'ON' at different strengths, or it may have different functions depending on where or when it is active. This fact goes some of the way to explaining why we can't understand our own complexity with reference only to the 22,000 genes present in our genomes (far fewer than had been assumed necessary to describe how humans work). It is also why mutations in the same gene can cause independent diseases in the eye and the kidney. How and when a particular gene is controlled can mark profoundly different functions in entirely different tissues.

There are so many confounding and complicating factors in biology that the underlying logic is lost in the messiness of real cellular life: unpredictable variation set among unfathomable sophistication. A genome, with its full set of biological instructions, is akin to the score of a symphony. The full beauty is expressed in a performance, complete with interpretation and nuance that is not merely encoded in the quavers, crotchets and minims written on the page.

We sometimes call the seemingly inscrutable complexity of a living system not music but 'noise'. It's a combination of all of the things that we can't quite account for: things like natural variation in the behaviours of genes or proteins or molecular interactions that haven't yet been described. The natural progression of biology as a science to synthetic biology has meant confronting the nuances of this mess and scrutinizing it. At the heart of synthetic biology is the desire to avoid that mess by creating new life forms whose circuitry and programming is clear, simple and, crucially, built not for survival but for purpose.

Sniper Circuits

'Imagine a program, a piece of DNA, that goes into a cell and says, "If cancer, then make a protein that kills the cancer cell; if not, just go away." That's a kind of program that we're able to write and implement and test in living cells right now.'

Those are the words of Ron Weiss, one of the founders of the modern engineering discipline of synthetic biology and now a professor at Massachusetts Institute for Technology. He is describing a landmark study his team published in autumn 2011. By using the logic and language of computer circuits combined with biological components, they had built a tool that effectively acted as a cancer assassin.

The terminator circuit is an assembly of DNA components, built to serve a singular mission: to identify and kill a type of cancer cell. Once built, the circuit slots into the genetic code of a virus, itself modified to grant us control over its natural tendencies. When introduced to malignant cells it infects them and, just like many viruses do, adds its synthetic genome (including the assassin program) to the host's DNA. Obliging the natural trickery of virus infections, that host cell unwittingly decodes the killer circuit, and enacts the program that will bring about its own downfall.

The assassin circuit represents a set of five questions, each one seeking the presence of a particular molecule that is unique to only this type of cancer cell. If any of the answers to this molecular interrogation is negative, the program stops, shuts down and decays, and the cell lives out its normal life. If, however, the answer to all five identification questions is 'yes', termination ensues. The code in the circuit contains a gene that triggers the host cell's own inbuilt suicide program. It is simultaneously a conscientious and persuasive assassin, one that makes forensically sure its quarry is indeed the target, and then quietly asks the victim to kill itself.

The genetic circuit is complex but logical. While it may seem

trite to draw such a literal analogy with microcircuitry, this system has in fact specific logic calculations built into it that are entirely derived from electronics. It uses exactly the simplified processing logic of electrical components such as AND gates. It also uses NOT gates, which flip the input from 'ON' to 'OFF' and vice versa.

This particular malignant cell is perhaps the most well-studied laboratory cancer in the world, the HeLa cell. It is an undying ancestry of cells that is derived from the cervical cancer of a young black woman called Henrietta Lacks. A scrape taken from the cancer on her cervix was grown in a lab in 1951, and was soon identified as being 'immortal'. Normal cells age, grow weary, and ultimately lose their ability to divide. Cell death follows. But HeLa cells reproduce indefinitely due to idiosyncratic faults in their own genetic logic boards. The tenacious immortality of HeLa cells has meant that they are the most studied and scrutinized malignant cells in existence, having been shared and spread among labs throughout the world. With such a well-studied fingerprint it makes them a worthy testing ground for such a heavily engineered genetic circuit. Most cells are not known in such forensic detail, so foolproof identification would not yet be possible. But for HeLa cells, the five molecular questions allow precise recognition, so thorough that it qualifies as what is known in electronics as a TRUE value. The precision of the assassin circuit is such that is in effect a biological computer.

This system is a high-water mark in the nascent field of synthetic biology. As with all potential therapies, it has a long way to go before being available to humans. Currently, it has only been tested in cells in dishes. Animal tests will come next, where the degrees of control will be challenged further by the more chaotic and dynamic noise of a working creature.* As a therapy, though,

---

* And indeed from cell to cell. One of the confounding things about cancers is that their DNA mutates quickly, and causes highly variable responses to therapeutic attacks of this precision.

its accuracy is potentially shocking. Cancer being a multitude of diseases, and being prone to mutating as it spreads and grows, means that you're aiming at an ever-moving target. Chemotherapy and radiotherapies remain the most effective attack on tumours, though they are both damaging to malignant and healthy cells indiscriminately. Compared with the carpet-bombing and collateral damage of radiotherapy, this cancer treatment is a sniper.

It represents the immersion that researchers have made in a decade into the realms of programmable biological machinery. As a weapon against a disease, the function of the assassin program is clear, and its assembly is premeditated.

Clocks and Waves

The first forays into the creation of biological components occurred in 2000 with two papers in the journal *Nature* merrily breaching the barrier between biology and electronics. One was the creation of a biological clock in the bacteria *E. coli*. By piecing together three sections of DNA that naturally prevent other genes from producing their proteins, Michael Elowitz and Stanislas Leibler at Princeton University constructed a circuit that dictates not merely that a gene is ON or OFF, but that it is ON and OFF in an oscillating wave. The wave continues as each gene inverts the output of the next: one switches the next off, which switches the next on, and so on. Cells have a natural cycle as they reproduce and grow, but this circuit was unlinked to that. The output of the circuit was the expression of the gene for Green Fluorescent Protein, and so the cells would pulse green in slow waves.

The second component part was made by a team from Boston University led by Timothy Gardner. They built a genetic version of a type of electric component known as a 'bistable', or more evocatively a 'flip-flop'. These types of switches flip between two states, but each of these positions has a function. You push the button on a CD player to turn it on, and again to turn it off, but in the

off position, the player is actually on standby, rather than cut off from the mains supply. It has two states, ON and ON STANDBY, and flipping between the two is fully reversible.

Both these engineering projects not only adapted the language of life to build unnatural gadgets, they also adopted the language of electronics to convey the functional design ethos these gadgets were bringing to biology. They were given names that could be straight out of the lexicon of electronics (or possibly science fiction): the glowing wave circuit gained a new electronics-sounding moniker – the 'repressilator', and the bistable flip-flop device borrowed from an existing electrical part: a 'toggle switch'.

These two inventions are rightly seen as the first synthetic biology parts: microscopic tools designed to enact a program, but built from DNA. The synthetic biology workshop opened with these first parts built, and what followed was a cascade of other mechanisms, tools, parts and pieces, all made out of DNA, taken, modified, fused and redesigned from evolution's toolbox. In the last ten years we have gone from the species-barrier-breaking insertions of genetic engineering to an ever-expanding toolbox packed with widgets. There are now switches, pulse generators, timers, oscillators, counters and logic calculators. The combinations of these and many other components has meant that our control of living systems extends to genes, protein function, cell growth and reproduction, metabolism and the ways cells talk to each other.

Less obviously practical than the cancer assassin program, but no less complicated, was a circuit built in 2009 by Jeff Hasty and his colleagues at the University of California. Timekeeping cycles are fundamental to most living things. They are known as circadian rhythms, and determine all sorts of patterns of behaviour in relation to the flow of time through our lives, such as the passing of day and night. We rely on cells to regulate our own metabolisms, such as in the rhythmic release of insulin, or in the orchestration of sleep cycles, both of which are disrupted in people who regularly work night shifts. How these timers work is partially understood, but they use similar mechanisms to the artificial

biological clocks built in these synthetic circuits. Hasty wanted to design synthetic circuits that created a regular pulse of activity, like the ticking of a clock. These so-called oscillators work as cellular timekeepers, determined by the careful design of a loop of genetic activity within a bacteria, three genes following a circular set of instructions. Imagine three friends standing on the corners of a triangle, twenty yards apart, all facing inwards. Each has a chair, but they only stand if the friend to their right is seated. When the first stands, the second sits, and the third stands. The first then sits, and the second stands, and so on. This continuous response is called a negative feedback loop, and the reaction time of the three creates a clock of activity. It won't work with four people, as the flow will end as the fourth person will be sitting after one round, and that will not prompt the first to sit down again.

As it is, this circuit will continue as long as the people can manage it. There is no output, though. Now imagine that, as well as standing and sitting, the first chap alone also has to sing a note, but only while standing. As long as the game continues along those rules, he will be singing every other complete circuit. If you close your eyes and only observe the output, you would hear a regular burst of song. This is a basic biological oscillator.

In a synthetic biology oscillator, when gene A is active, it shuts down gene B, which activates gene C, which completes the circuit by shutting down gene A. The program is constructed from those three component parts – genes and their activation sequences – and imported into the guts of a bacteria. In the version involving people, the frequency of the burst of song is determined by the speed at which each of the friends reacts to their neighbour. In the cell, it is the speed at which the protein is produced. Gene A is translated into a working protein, which binds to the on/off switch for Gene B, and so on, but each step takes time. The clock function is a result of this delay. As with the note being sung, the output of the circuit in bacteria might be the expression of a fluorescent protein, which will slowly come on and off with a regular pulse. In

synthetic biology, the aim is to elicit control from these DNA circuits, so the ability to fine-tune is desirable. Adding other loops and other mechanisms can introduce further control to the frequency and strength of the output, adding to the complexity of the logic.

One limitation of this basic biological clock is that the simple oscillator circuit is self-contained. In clock terms, this would be as if all wristwatches kept time perfectly, but everyone's was set to a different time-zone. Time is most useful as a commodity only if we all synchronize our watches, otherwise watching the *Ten O'clock News* would be a lottery. In bacteria, engineering a single cell to produce a regular pulse is one thing, but to turn that pulse into a synchronized wave in a population of bacteria is wholly more difficult. *E. coli* reproduce every twenty minutes or so, so it's a busy crowd to exert military control over.

In Jeff Hasty's program, though, the circuit also sends out a signal to all neighbours, which syncs the beginning of the loop. Rather than three people in a triangle, it becomes a Mexican wave in a football stadium. You still stand when your neighbour sits, but it happens in banks of people at the same time. The pulse, the output of a hearty cheer, is bellowed not by one person but by thousands. The bacteria, speeded up in time-lapse video, pulse with glowing green, like a luminous ripple. Anyone who has ever been part of a decent Mexican wave will know that it's impressive enough. But the population size of these bacterial populations is far greater than any stadium capacity. The act of inducing perfect synchronicity in these synthetic bacteria is more like making every traffic light in the world flick to green at exactly the same time.

That's a neat trick indeed, and shows the level of control these new synthetic circuits give us. But it also has valuable potential as a tool for synchronizing the output of a population of synthetic cells.

Insulin's job is to tell the body how and when to process carbohydrates and fats after you've eaten some roast potatoes. Glucose is the fuel that powers cells (and by extension, organisms), but the

concentration of this simple sugar in your blood is a delicate balance: too much or too little is deadly poisonous. We take in glucose directly via our diets, and other foodstuffs get converted to glucose in our muscles and liver, where it is stored in other forms such as fat. We have evolved all sorts of mechanisms and biological cheat codes to make sure that our cells are fuelled by a constant supply of glucose, regardless of whether we have just scoffed a chocolate bar or not eaten for hours. Insulin is a hormone that is integral to maintaining glucose levels, and it does so by instructing cells in various tissues to take up glucose from the bloodstream. In doing so, glucose levels in the blood are reduced, and the conversion of glucose from stored fat is turned off. The production of insulin is prompted by high blood glucose, and shut down when it reaches a threshold, so is in itself part of a feedback loop. Importantly, the curiousness of insulin production in the body is that, even when we are at rest, the level of insulin is not still. It oscillates gently at a rate of between three and six minutes regardless of blood glucose, like an idling car. When this process malfunctions, diabetes is the result. A hypothetical, synthetic circuit that can oscillate in its production of insulin (or anything) in a whole population of cells, rather than in an individual, would be of enormous benefit in mimicking the natural pulse, and ultimately a potential therapy for diabetes patients.

As with Ron Weiss' cancer assassin program, these engineered biological circuits are not nearly ready for clinical use. That program is carried in a modified virus, unable to reproduce and unable to enact its software without the machinery of existing living cells. The virus has to infect the cell before it can enact its diagnosis and sentence. But most of the circuits of synthetic biology are integrated into the genomes of bacteria. We've not reached the point where we can test synthetic circuitry in humans. And a long time before we get to that point, there is the not insignificant problem of delivering the software and its packaging to the place where it's needed. The bacteria themselves are merely vessels that carry the software and provide the mechanics of producing that

circuit's output. Humans are covered in bacteria and full of them. They are much smaller than human cells, and so we can carry maybe ten times more bacterial cells than our own. Almost all are benign or beneficial, especially in our guts, where billions of bacteria, our microbiome, provide digestive functions that we don't inherently possess. But our bodies are good at detecting and destroying invasive species, so a genetically enhanced bacteria would need to be undetectable to the immune system, coated in a molecular invisibility cloak such that the legions of intruder alarm cells that patrol our bodies simply fail to identify the synthetic cell. But our immune systems have had billions of years to evolve the ability to detect foreign invaders, and free-roaming synthetic bacteria would be policed harshly. So, an alternative would be to package the bacteria up and hide them from the immune system. Unlikely though it sounds, NASA are working on precisely that problem.

Under the Skin

NASA's base at Ames resembles a small American everytown, just off the Californian freeway, with wide roads, grassy squares, faux stucco buildings and acres of blue blue sky. Except that, in between these blocks of normal, there are colossal paeans to the grandstanding 1950s optimism that science would deliver the future. The biggest wind tunnel in the world occupies several blocks, its interior roomy enough to accommodate a full-size aeroplane. The air intake is a giant black rectangular grille so large that it's difficult to gauge its colossal scale, until I spy some NASA scientists playing street hockey at its base. And then four blocks to the north is a gargantuan hangar built in the 1930s to house the building of dirigibles, the short-lived rigid airships of the past. In a clean-room round the corner, the next moon-bound spaceship – called LADEE – hovers as its carbon-dioxide-based levitating propulsion engine is tested by technicians.

NASA, of course, exists to explore space. On the corner of a spacious crossroads there is a building whose concrete façade is pockmarked with casts from meteorite impacts. Inside, the researchers in NASA's synthetic biology programme are using the microscopically small to solve some of the biggest problems that lie between us and our desire to explore strange new worlds.

The Sun beats down at Ames; whilst being the bringer of energy and life on Earth, it is far less charitable off-world. Sporadically and unpredictably, our star will blast out a shot of extremely high-energy particles with enough clout to disrupt power supplies on Earth. Similarly, other stars in the universe eject a continuous bombardment of high-energy particles from deep space that bathes the galaxy in radiation. Beyond the protective blanket of our atmosphere, solar flares and galactic cosmic rays combine to be one of the most restrictive dangers for human space exploration. Radiation, including in these interstellar forms, has the effect of slicing up DNA randomly. Often that is not a problem: DNA repair is a major industry within our cells, with legions of copy-editing proteins beavering away to make sure that errors in the code, or mismatches in the two strands of the double helix, are identified and patched up. Occasionally the damage may occur in one of the many genes that control cell division, and if that happens in a gene whose purpose is to instruct a cell to stop duplicating itself, then you have the beginnings of a cancer – uncontrolled and unrestricted cellular growth. Conversely, breaking up the many genes that are necessary for the cell's continued existence can result in the inception of a natural cell suicide program.* Unrestrained cell death or cell growth are equally bad for an organism's wellbeing. We are constantly exposed to damaging radiation, but mostly in small doses or infrequently. This is why, when you have an X-ray in a hospital, the technician will stand behind a protective screen.

* Cell death is, perversely, essential for growth and a very important part of development in the womb. For example, as foetuses our hands start out as paddles, and the fingers grow within them. The fingers separate when the cells in the webbing enact their own death.

It's a rare event for you, but unshielded, they are exposed dozens of times every day, with potentially deadly consequences. In space, the exposure is continuous.

In 2005 the US Federal Aviation Administration set off on a virtual round trip to Mars, including a fourteen-month stopover on the red planet. They calculated the amount of exposure to space radiation astronauts would receive, and deduced that astronauts would suffer a significantly increased lifetime risk of cancer (as well as a host of other conditions, including cataracts and sterility) as a result of exposure to cosmic rays and a bout of radiation from a solar flare. The safe limit was a journey of around fifty million miles, that is, just short of halfway to the Sun. Shielding on the spacecraft prevents radiation exposure to humans, but it is heavy, and weight is a critical issue when calculating the propulsion and cost of getting off our planet. The hostility of space means that our biology is just as limiting as our engineering when it comes to exploring the heavens.

Our immune systems do have inbuilt defences that counter the effects of radiation damage. We have these mechanics in place to deal with the much weaker radiation – ultraviolet – that causes sunburn and other forms of DNA damage. When detected, the severed DNA switches on genes that produce small molecules called cytokines, which flit from cell to cell, triggering cascades of repair programs. But they have limited reach, especially in reply to prolonged exposure to damaging radiation, and this is why cytokine therapies are sometimes used to enhance the body's natural response.

David Loftus and his team at NASA Ames are planning a designed synthetic program in which the input is a detection system for radiation or DNA damage, and the output is the controlled release of cytokines. This program is different from the cancer assassin in that the output is a molecule that stimulates the body's own natural immune system. So carrying the program in a virus that will infect a cell will not work. The production needs to be self-contained. Therefore a bacteria is the most appropriate synthetic

cellular manufactory. The question is how to introduce a bacteria safely into humans without provoking an immune response.

The circuit is not yet built, and this construction will face all of the same problems that any synthetic biology project does. But Loftus and his team have made stunning advances in the construction of the capsule to contain the cells. As if the circuitry of synthetic biology was not impressive enough, the biocapsule is made from carbon nanofibres. A minuscule mould sits on top of a hypodermic needle, which is attached to a pump. This is submerged into a suspension containing the carbon nanofibres, and they are literally sucked onto the mould to form a capsule. This tiny pellet is half a centimetre long and half a millimetre wide, not much bigger than this letter 'l'. Although the capsule is tiny, bacteria are much smaller, so it is large enough to imprison tens of thousands of them.

It is also biologically neutral, meaning that the carbon nanofibres will not provoke an immune response and will cause no harm to the astronaut. But with serendipitous grace they bunch together in a porous tangled mesh like a sheaf of frozen worms visible only under electron microscopy. The gaps running through this mesh are too small to allow bacteria through, but big enough to let small molecules, such as the cytokines that tackle radiation damage, seep out. The idea is simple enough. The synthetic bacteria cells are placed within the capsule, and the capsule is implanted under the skin of the astronauts. The cells contain a program that produces and releases cytokines upon exposure to solar and cosmic radiation. It requires no diagnosis or intervention, and, elegantly, the treatment for the sickness is prompted by its cause.

Imagine this technology applied to less exotic practical human problems: a synthetic cellular circuit, permanently implanted under the skin to deliver the treatment to a disease without the patient ever knowing. The potential is breathtaking. Diabetes sufferers stand out as being obvious benefactors. The cells in the pancreas that produce insulin could effectively be replaced with synthetic organisms that serve exactly the same purpose. Instead

of injecting insulin, a synthetic circuit borne by bacteria hidden in a biocapsule shell would produce insulin according to the body's fluctuating needs. In principle, the patient would never even be aware that this process was going on.

As ever with brand new and untested inventions, this exact mechanism may yet come to naught. It may turn out to be impractical, or too expensive to be of general use. At best, this system is a decade away: the capsule is still being developed, and the synthetic circuits themselves are years from being tested in animals, and therefore years from being tested in humans. But this shows how the need for practical solutions to complex problems typifies the renaissance nature of synthetic biology.

Reality Check

What I've described above is the promise and the hope of this new engineering project. I've outlined some of the highlights of ambition and success of this emerging endeavour. Currently, there aren't that many more, though the volume of synthetic biology publications is rapidly increasing. It is easy to get swept up in the hype as the rigour of the lab is translated into terms aimed at non-experts designed to inform, enthuse, or generate criticism. That said, the comparison with electrical engineering is one that has been made by synthetic biologists themselves, in an attempt to standardize the parts of DNA into component parts in order to construct living machines.

But the reality is that referring to living things as machines glosses over the facts of life. Cells and organisms are machines whose complexity is many orders of magnitude greater than an engine, or a production line, or even a computer. Engineering, like evolution, is an iterative process. The incomprehensibly deep time since the origin of life on Earth has meant that the number of tests and failures and rebuilds and further tests undergone by the mechanisms of life is impossibly large. Every organism that has

ever existed was an iteration in a perpetual test of engineering function, unguided and ruthless: will it live, and live to reproduce itself? This is true, of course, of any species or organism, but the real selection under scrutiny in evolution is individual genes, the functional natural component parts of living things. These work in consort, in networks, in cascades and in meandering dynamic pathways. The complexity of a living organism has billions of years of successful testing behind it, and that means that to simplify or design genetic circuits into simple (or even complicated) tools is tricky. Biotechnology expert Rob Carlson told *Nature* in 2009 that 'there are very few molecular operations that you understand in the way that you understand a wrench or a screwdriver or a transistor'.

The test of design in engineering is 'does it work?', and more precisely, 'does it work as we designed it to?' In this new discipline, the answer for many of the several thousand parts and circuits currently available is a simple 'no'. The complexity of biology in cells generates noise – unpredictable variation that masks or subverts the intended output. The so-called repressilator that marked the beginning of synthetic biology in 2001 by making bacteria blink glowing green worked impressively well. But the surprise was that not all cells worked the same. The pulse of the glow was nowhere near consistent: some cells were brighter than others, some slower, and some skipped a beat. The reasons why are not impossible to understand, but complexity breeds inscrutability, and that makes the reductionist ethos of engineering harder to fulfil. Once again, the analogy with electrical engineering even holds true here. We already make digital logic boards so dense and so filled with software that we don't fully understand their behaviours. This is why computers crash. These systems are built for a purpose and tested and retested to make sure they succeed at the tasks we design them for. But that does not mean that we know or can know all potential behaviours that they may express. In computer hardware design, failure analysis is critical, just as it has been in human genetics for a century. We have always discovered genes

and their functions only after observing them fail, that is, when they cause disease. Well-designed hardware is built to cope with the unpredicted and unpredictable noise that inscrutable complexity may host.

The problems caused by mutated genes that result in disease have a clear origin, if you know where and how to look for them. In the wishfully simplified programs of synthetic biology, it is the unforced errors that are problematic. Decoding the obvious faulty mechanisms of diseases, and designing successful programs are certainly the aims of genetics and synthetic biology. But understanding noise and catering for it remains a puzzling obstacle. Because of this, so far, the parts, the circuits and the living hosts make the dream of a standardized DNA construction kit a way off yet. But this is a young field, immature and full of hope. Typically insightful, the doyen of twentieth-century science fiction Isaac Asimov noted that the best moments in science are not when you say 'eureka', but when you say 'hmmm, that's interesting'. That noise, the fact that these programs don't quite work as designed, is definitely a problem, but a most interesting one.

Nevertheless, parallels with computing are striking. The legends of the computer industry's roots describe future billionaires like Steve Jobs and Bill Gates tinkering with electronics in their garages, playing with components and code to make better hardware and better software. The results were Apple and Microsoft, a computer on every desktop, the Internet, Google and a smartphone in every pocket.

Whereas political revolutions tend to be defined by events – the storming of the Bastille, or the execution of a head of state – the cultural revolution of computing was not triggered by any single act. Similarly, workers in nineteenth-century Europe were aware that new-fangled technologies, such as looms and Spinning Jennies, were driving significant changes to their working lives, and those changes were happening at a pace. But the aggregation of that change and its significance were unknown at the time, and the term 'industrial revolution' was introduced much later. The

practitioners of synthetic biology are creating in a way that is novel for biology and for science, and all within a period of a decade. This manipulation of four billion years of evolution specifically for the creation of unnatural biological tools is potentially a revolution occurring right now.

Richard Feynman's aphorism is crucial: to understand something you must first be able to build it. The terms, the language used and the techniques are all borrowed from the problem-solving ethos of engineers, a reductionist view of function through design. The extension of natural processes to serve humankind's desires is not new, but has rarely been approached with such a clear sense of construction built in. Though the founders of synthetic biology had electrical engineering in mind, the metaphor that has stuck most has been a toy. Just like DNA, Lego is universally adaptable too, with the pieces designed to fit together regardless of whether they came from a spaceship or a castle set. Similarly, synthetic biologists have tried to make the individual components of their circuits interchangeable, so that construction of novel circuits is not hindered by the natural noise of biology. With that in mind, the spirit of invention, engineering and creativity has been built into synthetic biology in a unique way. But as this revolution unfolds, questions of ownership, legality and ethics press urgently upon us.

3.   Remix and Revolution

'Rien ne se perd, rien ne se crée, tout se transforme.'

Antoine-Laurent Lavoisier, *Elements of Chemistry*

Evolution is the most creative enterprise that has ever existed. Nothing comes close to the diversity, sophistication and beauty that has transpired at the hands of DNA and natural selection. At its heart are two words that carry a potentially derogatory sense: imperfect copying. In another sense, evolution is the ultimate exemplar of that maxim made famous by Isaac Newton – 'if I have seen further it was by standing on the shoulders of giants' – which neatly describes the derivative nature of creativity, while exemplifying it, as he was borrowing and adapting the same idea from twelfth-century philosopher Bernard of Chartres, who was in turn probably referring to ancient Greek versions of the same idea. The French chemist Antoine-Laurent Lavoisier's 1789 maxim at the top of this page was a comment on the nature of matter, but works equally well for energy, biology and ideas: 'Nothing is lost, nothing is created, everything is transformed.'

It's easy to apply a similar principle to culture, that, while there may be nothing completely new under the sun, the creation of new ideas is still born of copying, adapting and transforming what has come before. In music, for example, it's not too difficult to trace an evolutionary path of influence through several centuries of creativity from Bach to Haydn to Mozart to Beethoven to Mendelssohn. Or, in modern times, from Chuck Berry to the Beatles to Bowie to punk to Joy Division to the Smiths to the Stone Roses and on and on.

The nature of copying in music changed in the 1960s when a

new method of musical creation emerged that was unlike anything
that had come before. Technology enabled musicians not merely
to copy and modify what had come before, but to directly borrow,
co-opt and lift it. The sampler allowed music producers to take the
drums from one existing track, the horn section from another and
the vocals from a third and mix them together with any other elements to create a new sound.* As a technique for creating new
sounds, sampling really took off on the streets of New York with
the emerging hip-hop scene of the 1970s. DJs in the Bronx would
simultaneously entertain crowds, pay homage to their musical
forefathers and create radical new sounds by mixing records
together on twin turntables. They would take the riff or beat from
one and drop rhythm or lyrics on top from others, often using soul
records as their licks. In live performances at clubs, and later in
recordings, they were not playing new music but creating it by
adapting and remixing existing sounds. By the early 1980s, sampling machines were being used to record short extracts from a
record so that it could be integrated into a new tune. It could be
slowed down, speeded up, looped or stretched to help create a
brand new sound that was derived from and built on the shoulders
of previous works.

Synthetic biology is remixing. The ethos of this emergent scene
is one of unprecedented and unbridled creativity that is a remix
culture. DJs did not have to be virtuoso players of specific instruments in order to create new sounds. In synthetic biology, the
creators do not need to be geneticists, DNA technicians or even
biologists to build new organisms. The principle is simply to create.

From inception many of the pioneers of synthetic biology have
striven to build a movement of open democratic science, where
there is a free exchange of ideas, techniques and materials. If Year
Zero of this new industrial revolution was the repressilator and

---

* The Beatles did this as early as 1967 by borrowing and looping calliope record-
ings on 'Being for the Benefit of Mr Kite' on *Sergeant Pepper's Lonely Hearts Club
Band* (and even more so on experimental tracks such as 'Revolution 9' on the
White Album.)

the toggle switch in 2001, the first big movements happened a couple of years later. The transition from tinkering in labs to a movement in science occurred over the course of 2003 and 2004, with the emergence of two phenomena that would become emblematic of the spirit of synthetic biology.

One was the founding of the Registry of Standard Biological Parts. For all the talk of the wonder of the interchangeable genetic alphabet, it takes a lot of code to make a life form. That means that there is a lot to play around with, and more importantly to get wrong when remixing your code. Ironically, for something to be open and free for everyone else to use, it needs to be codified and standardized. Think how annoying it is that electrical plugs are different in different countries, or phone chargers change from one model to the next. The language of music is universal, but you need a whole battery of adapters and plugs to perform it in different countries. But nuts and bolts are standardized, so that thread size or screw dimensions don't have to be reinvented every time you want to assemble a piece of furniture.

Transfer that to the unlimited amount of coded information contained in the new circuits of synthetic biological components. Dozens, and then hundreds, of labs started building their own gadgets and devices, all of which only work in circuits built into genomes of living things (or, in some cases, near-living viruses). Without standardization, useful sharing is hamstrung. So in 2003, the BioBrick was conceived by Drew Endy at Stanford, Tom Knight (then at MIT) and Christopher Voigt at the University of California, San Francisco (UCSF). The BioBrick is a useful and simplified way of plugging in bits and components so that everyone who signs up doesn't have to redesign each idea and each part for themselves. BioBricks are to genetics what the sampler is to music – a system for freeing up the elements of the rich biological past and the ingenuity of others' designs, in order to foster extreme forms of creativity, by standardizing the assembly of DNA components so that each link doesn't need redesigning every time. This is where the comparison with Lego holds some weight. Lego

bricks are beautifully designed to connect with any other Lego brick. There is a universal fit to the pieces, and they can be assembled regardless of their original kit. At the time of writing, the BioBrick catalogue contains more than 10,000 parts. Each one is a piece of DNA, and is delivered in the mail as a dot on some blotting paper. Drop it into solution, and the DNA floats off the paper, ready for assembly. Some BioBricks are genes, some are regulatory instructions, and some are combinations of both, already assembled to be integrated into new circuits. All have been standardized so they can be knitted together for new creations.

The other transformation that occurred at this time was the inception of the International Genetically Engineered Machine (iGEM) competition in 2003 and 2004. Each year, teams of undergraduate students devise a problem, and design and build their proposed solution using only the components available in the Registry of Standard Biological Parts. Students compete to get into their college teams, and spend weeks researching, designing and enacting their solutions to their chosen problems. It is a talent competition for the brainy, and the excitement is palpable. Each team receives a kit consisting of a thousand BioBricks from the Registry.

After rounds of regional selection, the finalists meet up at MIT in November for the annual jamboree. Here, the teams show off and celebrate their creations, and synthetic biology pioneers and leaders decide a range of prizes. The Grand Prize is emblematic of the construction mindset: an oversized aluminium Lego brick.★

The democratic principle is built into the Registry and the iGEM competition too. The resources are all free, but the expectation is that all players will contribute. The Registry website asks that users 'get some, give some'. All teams compile their notes and projects, successes and failures on open wiki pages, designed to encourage sharing and standing on the shoulders of peers.

★ This was replaced for the 2012 competition by an elegant wooden giant Lego brick.

Although few of the iGEM projects have come to fruition yet, there is a vibrancy about this endeavour that is unlike any science meeting I have seen before. Many undergraduate students take on summer projects in the laboratories of their tutors. Here, they often do menial or slavish jobs, part of the grind of data production that science is based upon, and in doing so gain experience of the reality of lab life. But iGEM teams create new and potentially important projects and solutions with a zest that is thrilling.* They spend their summers inventing solutions to global problems. They are combining brand new, often untested, tools, and in doing so are building new ones, using the most advanced biotechnology available. They then present their work to the judging panel, which is made up of a hefty proportion of the most significant synthetic biologists working today.

Hello World

In 2004, a striking entry to iGEM came from a team-up between University of Texas, Austin, and UCSF. They designed a circuit and successfully integrated it into bacteria, which then acted as a photographic plate. One component was a light-sensitive protein, which would perform its function when stimulated by a beam of light. This they linked to a standard pigment-producing gene called LacZ. When they grew the synthetic bacteria on a flat clear plate and shone an image onto it, the bacteria processed the light and dark at the resolution of 100 megapixels per square inch. The work went on to be published in the journal *Nature*, and the components developed into a new synthetic biology device: an edge detector. The projected image was a text message, appropriately

---

*In the summer of my second undergraduate year in 1995, I spent three months measuring the vital statistics of 3,000 stalk-eyed flies to assess whether asymmetry was greater in their eye-stalks compared to their wings. It was not. The world of biology remains unquaked by this study, but the lure of the lab was strong enough for me to return.

enough announcing the arrival of this new field, but also an in-joke displaying synthetic biology's computing roots. The phrase is the output of a standard test that programmers use to check their coding language is functioning correctly. It simply said: 'Hello World.'

The Grand-Prize-winners in 2009 were from Cambridge University: they extended the use of glowing jellyfish proteins (GFP) and designed a system for bacteria to produce multicoloured dyes as a tuneable detection tool. These are known as biosensors, and they already exist, such as for detecting blood glucose levels. But every biosensor so far has to be designed from scratch and built specifically to detect just one target. The Cambridge iGEM team aimed to build a generic biosensor that could be tailored to multiple uses. So, for example, the engineered bacteria could be tuned and used as a detector of a specific toxic chemical in the environment, and would produce a colour-coded pigment in response to its concentration. They also eschewed the use of GFP, as this requires fluorescence kit to see it. Their aim was to make a biosensor that could be seen with the naked eye. The parts of the circuit are the sensor, a sensitivity tuner and the pigment generator, six of which are new parts, and were submitted back into the Registry for others to use.

Washington University took the main prize in 2011 with a circuit that produces forms of fossil fuels. The runners-up, from Imperial College London, have attempted to fight desertification – the gradual erosion of fertile ground into agriculturally and therefore economically redundant land. Some estimates predict that, by 2025, as much as two-thirds of Africa's arable land will be rendered barren by this phenomenon. The Imperial team has engineered genetic circuits in cells that will nurture roots, which in turn shields fertile topsoil from the eroding elements. Seeds are coated in the synthetic bacteria and, once they germinate, the cells inveigle their way into the infant roots, and the program they bear produces the plant hormone auxin, which accelerates growth into the topsoil.

The Slovenian national team appear to win the iGEM Grand Prize when the year is an even number, claiming the prize brick in

2006, 2008 and most recently in 2010. That year, they took the principle of using biology for purpose into new territory by eschewing the language of DNA altogether. Instead of manipulating the code embedded in the A, T, C and G bases of DNA to create new function in translation, they used the simple fact that we can put those letters in a specific order to build a miniature production line. Living things are full of such production lines: one gene produces a protein that triggers another reaction, that triggers another gene and so on. But, unlike the ordered machinery of a factory, the inside of a cell is a busy milieu, with all the code and proteins effectively free-floating in the gooey plasma. Real factories would be disastrously unproductive if each part was allowed to randomly drift to the next step on the production line. Slovenia's team designed a sequence of DNA that organized the unsecured mess into a neat line, by using DNA as a production-line scaffold. Instead of having proteins free-floating in the cell, they figured that they could tether them to a unique sequence of DNA. Many proteins bind to DNA anyway, so much is known about how to get a protein to clamp onto a specific sequence of DNA. With the right stretch of DNA each protein in a pathway can be pulled out of the cellular soup and tethered sequentially, resulting in a biological production line. The primary aim is that this scaffold will radically increase the efficiency of synthetic biology pathways. But it also opens up the possibility of more tightly controlled oscillator switches or other standardized devices in the synthetic biology toolkit.

Careful Packing

Almost all of the iGEM teams come from universities. But, as described in the previous chapter, NASA has been quick to identify synthetic biology as a discipline that will enable their continuing mission: to boldly go. Among the more fanciful ideas that NASA is exploring are their plans for synthetic bacteria that make bricks.

This challenge was taken up by students in the iGEM

competition in 2011. An hour from San Francisco, Stanford University collaborated with NASA Ames, just a hop over the freeway, and Brown University to figure out how synthetic biology could help us build colonies on other planets. Led by Lynn Rothschild, they have asked how synthetic circuits can contribute to terraforming. As mentioned earlier, the issue of weight is absolutely critical in space travel. It is estimated that the act of launching a thing beyond the Earth's gravitational pull costs around $10,000 per kilogram. The mass of Saturn V, carrying all the modules to get the three crew members of Apollo 11 to the moon, have a stroll for a few hours and come back home, was around 3,000 tonnes. The lunar module itself, the Eagle that landed, was 17 tonnes. Estimates of a round trip to Mars include consumables and life support of 200 tonnes alone. So rather like packing for holidays, minimizing what you take with you and using what you find when you get there is the best way to do it. 'In Situ Resource Utilization' is NASA's official term, but I prefer to call it 'careful packing'.

Given that our dreams of establishing permanent bases on other worlds are profoundly hindered by having to carry the building materials from Earth, using the materials on another planet is appealing. The question is 'can we build with moondust?'*

The Brown–Stanford iGEM team founded their project on this question. They happened upon the alkali-loving bacteria

---

* In the 1980s, a team of lunar mechanics got to play with a huge amount of regolith, the powdery dust that makes up the Moon's surface. A huge amount, in this case, is 40 g, about the equivalent of four tablespoons of sugar. This precious sample, brought home by the astronauts of Apollo 16, was cemented into tiny bricks, an inch-by-inch cube and a couple of other shapes, and was tested to destruction. The researchers concluded that 'Lunar soil can be used as a fine aggregate for concrete.' One of the key ingredients of concrete, however, is water, and though the moon is not entirely desiccated there isn't a lot of water there. In 2009, the spaceship LCROSS was deliberately crashed into the Cabeus Crater, a pit shrouded in permanent darkness on the south side of the Moon. NASA very carefully watched the 2 km plume that erupted after impact and saw from its unique ultraviolet signature that water and ice puffed out. It's buried, though, and not available in quantities that would enable construction of a moonbase.

Sporosarcina pasteurii, whose metabolism involves expelling the mineral calcite as a waste product. In the right type of sand this undergoes a process called biocementation, that is, it forms cement. Water in normal concrete forms a glue by a range of chemical reactions, and is used up in the process. The cells that secrete calcite can serve this purpose without water, other than what they need to live. The key component of this process is a protein called urease, which initiates the breakdown of urea, and in these bacteria ends with cementation outside the cell. The iGEM team identified the genetic circuit for urease production in *Sporosarcina*, and copied it into a standard lab cell, *E. coli*. They then characterized and standardized it into a BioBrick, where it became known as Part: BBa_K656013. When expressed in a cell, and placed in regolith, the circuit is activated and has the ability to cement it into bricks. They have tested this process using simulated Mars sand, admittedly in moulds just a centimetre across, but within a few hours they solidify into tiny bricks. It doesn't need saying that this is an astronomical distance from being used as a building material on Mars, but that journey is a decade away at best, and at the rate synthetic biology is progressing, BioBricks making actual bricks from Martian regolith becomes a whole lot more plausible. True to the playful roots of the BioBricks mentality, they are called RegoBricks. The weight saving is enormous. A small vial of these remixed cells weighing only a couple of grams could be transported aboard a ship, and grown into a RegoBrick factory on the planet itself.

These are just a handful of some of the hundreds of entries to the iGEM competition during the few years it has existed. The BioBricks Foundation aims not just to spread the potential of applying engineering principles to biology, but to do it in a way that fosters open, egalitarian science, creativity and a free exchange of creativity, much like the remix musicians in the 1970s and 80s.

While this may sound somewhat idyllic, there are problems with iGEM and BioBricks. In terms of bearing fruit, more often than not, iGEM teams do not actually deliver the products they have

designed, often due to the time constraints of the competition (though many individual parts and circuits have resulted in published scientific papers further down the line). But there are plenty of other issues with BioBricks. Hundreds, if not thousands of the parts in the Registry are not well characterized. The individual parts include genes as well as the instructions to turn genes on or off. But many of these behave inconsistently or unpredictably, and much of this may be to do with the biological noise of a living cell, as compared to the logic diagram of a design. The nature of the competition means that often students don't have time to fully characterize the parts they have built. Standardization is becoming a Sisyphean task, with the rate of submissions overwhelming the urgent task of characterizing what is already there.

Nevertheless, the excitement of the competition is electric. Teams contain students from backgrounds in maths and engineering as well as from the biological sciences, and bring a fresh ignorance to the problem solving. I went to the UK team meet-up in September 2012, organized by University of East Anglia's team leader, Richard Kelwick. There the UK entrants presented their projects to each other and a handful of synthetic biologists in anticipation of heading off to Amsterdam to be selected for the world finals. Out of eighty-five students entering from the UK, twenty-nine were not biologists. The entry level for some of the most sophisticated, creative and advanced genetic engineering is as open as it has ever been. Of the three teams that went through to the next round, University College London's was the most outlandish, literally: bacteria that would collect and consume the billions of tiny fragments of unbiodegraded discarded plastic that float in the oceans, with the intention of using this harvest to build a plastic island, recycled and claimed from the sea.*

---

*The 2012 Grand Prize winners were the iGEM team from University of Groningen in the Netherlands, who built a circuit into bacteria that detects chemicals that are produced when meat begins to go off. The bacteria then produces a pigment visible to the naked eye, so that consumers can see when meat is rotten.

This open access is deliberate and inherent in the Biobricks mentality: a profound understanding of how biology works is not necessary if you are to use that biology merely as tools. While this may seem at first glance questionable, it is how we operate with all technology. I type these words onto a computer whose hardware is largely mysterious to me, using software written in a language as foreign to me as Mandarin. Even the simplest mechanical tool, such as a monkey wrench, is made by a process that requires expertise unknown to most people who use them. With the application of engineering principles to biological components, the question is not how these parts work, but whether they work or not. This way, biology becomes non-exclusive. By standardizing the parts, the entry-level criteria are lowered past the point where traditional expertise in the ways of DNA are required. Paul Freemont runs Europe's largest synthetic biology department at Imperial College London (hosts of the bi-annual premier synthetic biology meeting in 2013, SB6.0), and he told the BBC in 2012 that they have 'a whole load of new people coming in, young, energetic researchers. They don't care where the biology is, they just want to build things [and] they want to solve problems. They don't have any baggage.'

Who Owns Your Creation?

As the purpose of synthetic biology is to create solutions for real-world problems, the potential for commercialization is obvious. So inevitably the question of ownership looms over any components, devices and products that get made. The story of patents and DNA is a tortuous and ongoing one, messy not least because the laws regarding patenting were not designed with biotechnology in mind. The promise that synthetic biology holds for future technologies means that the question of patents is intrinsically entwined with the science, and may yet have an effect on how this

necessarily creative industry will flourish. So it's worth having a mercifully brief look at the problems.

Patents are historically designed to protect useful functional inventions, whereas copyright protects the physical expression of an idea, be it in the form of a written or music manuscript, a photograph or a web page, by granting ownership, usually to the creator. But more fundamentally, both copyright law and patents were introduced in the US in order to foster creativity for the greater good. Inventors need protection over their creations as the time, money and intellectual investment they put into designing and making a new product is considerably greater than is required if simply copying it.

'Products of nature', on the other hand, are ineligible for patents as they occur without any human ingenuity. But it has been successfully argued that once a genetic sequence, such as a gene, is isolated and characterized, it is no longer a 'product of nature', as it has been copied and subjected to the intervention of humans. Therefore it can be protected by a patent. The most high-profile recent test case of DNA ownership was (and continues to be) for two breast and ovarian cancer genes, BRCA1 and BRCA2, both of which are owned by the company Myriad Genetics and the University of Utah Research Foundation. At the time of writing, this case has been bounced back to the Federal Supreme Court in the United States on the grounds that the techniques applied to isolate the DNA are standard molecular biology and not unique. The back and forth about DNA patents rumbles on. Nevertheless, primarily using the argument that the isolation of genetic sequences is a process, currently one in five human genes, the genes that you carry round in your cells, are effectively owned by someone else.

When it comes to patenting living things, the law, at least in the US, is marginally clearer. The first patent on an organism was upheld in 1980, with a genetically modified bacteria that helped crude oil break down. It was granted, on appeal, on the grounds

that its genes were at least partially man-made, and that the process of manufacture was patentable. The key phrases used in the ruling were that the invention (and therefore the patent) constitutes a 'manufacture' or 'composition of matter'. In 1988, this was extended to a multicellular organism: the 'OncoMouse' is a transgenic mouse whose DNA has been modified to make it particularly susceptible to various cancers, and therefore is a valuable research tool. And in 2010 Craig Venter filed a patent for the genome of his artificial cell Synthia.

As the products of synthetic biology have little historical precedent, it is not yet clear how the law will (or will evolve to) handle them. In this case the complexities of DNA patenting highlighted above are compounded by the fact that much of the thought and indeed behaviour behind synthetic biology components resembles, often by design, computer software. Here, it gets even murkier, as software is a field also beset by continual messy ownership debacles. Software falls somewhere between patent and copyright, as it is both functional (it enacts a program) and an intangible work (it is written to be enacted). The behaviour and use of individual parts or pieces of software source code, such as formulas and calculating machines, can change depending on how they are used. As a result, computing patents are sometimes bafflingly broad, encompassing very non-specific terms.

You'll forgive me wading into this dark world of intellectual property, but it has specific relevance to the nature of the BioBrick. As a result of the nebulous legality of biotechnology, broad patents have been used to grant intellectual property rights over mechanisms of cellular machinery. They cover general principles as well as specific processes. This has the effect of restricting experiment on those cellular activities, especially for researchers at the beginning of their careers, who cannot afford to pay for patented processes. If broad patents could be applied to the devices that synthetic biologists have created, such as over an oscillator in a generic sense, then it could potentially limit modification of what has become a basic instrument in the synthetic biology toolbox.

The BioBricks Foundation has recognized these potential pit-falls, and taken a very precise stance on this. They state that its mission is 'to ensure that the engineering of biology is conducted in an open and ethical manner to benefit all people and the planet. We believe fundamental scientific knowledge belongs to all of us and must be freely available for ethical, open innovation.' They have reflected that patenting and copyright law as applied to computing was riven with inadequacies. The BioBricks user agreement has enshrined an irrevocable protection from the assertion of intellectual property rights. Attribution to the creator and adherence to safety practices and, of course, the law are part of the agreement, but the part itself is quintessentially free to use. By placing their wares in the public domain, the users are precluding the possibility of thickets of patent enforcement, and consequently encouraging innovation and creativity.

In April 2012, recognizing the potential that synthetic biology has to change the world, President Obama launched the 'National Bioeconomy Blueprint', a roadmap of how to invest in synthetic biology in order to extract the maximum benefit from it. This plan has commercial development written into it, but also recognizes the value of BioBricks' core beliefs about sharing data and resources. Very specifically, this set of recommendations is designed to promote growth in synthetic biology, address safety issues, encourage commercialization without restricting creativity and ensure the public is coming along for the ride.

The protection of creativity is enshrined in the titles of patent and copyright legislation, both from 1790: the Copyright Act is subtitled 'An act for the encouragement of learning' and the Patent Act 'for to promote useful arts'. In music, computing, genetics and potentially now in synthetic biology, these principles have been mired in the wake of speedy technological advances that far out-stride changes in the law. As a result, in biotechnology patents are in danger of becoming the embodiment of the impediment of creativity. When that happens, progress halts.

Again, the parallels with music are striking. Sampling in

recorded pop music has become virtually ubiquitous. Hip-hop, which is based inherently on sampling, grew to become the biggest and most dominant business in music. As corporations recognized its immense commercial potential, the companies that owned copyrighted music also began to exercise and enforce control over the sampling on which hip-hop and other music genres were based. Copyright laws were refined and introduced that meant that, unless you have deep pockets, nowadays it is virtually impossible to make music with the same creative freedom that helped create and define this genre.

For synthetic biology, wrangling this tangled legal thicket is all part of the rapid maturation of a very young field. In 2011, one of its godfathers, Drew Endy, opened a new venture to provide support and an open legal framework for the continued development of biological devices, circuits and tools. At the launch he declared that '[w]e now need to move beyond Lego metaphors and genetic toys to professional technologies'. The message is that synthetic biology needs to mature, both scientifically and legally, and to become the global industry that its founders think it can be. But it should also continue to be fuelled by unbridled creativity and even playfulness, as this is the fertile ground from which the most innovative ideas grow. In the words of the Harvard law professor and copyright activist Larry Lessig, 'A culture free to borrow and build on the past is culturally richer than a controlled one.'

The creativity built into BioBricks, and synthetic biology in general, fosters an ethos that is distinct from the field of genetic modification that preceded it. It also has an engineering principle at its heart. Both of these aspects build up the case for seeing synthetic biology as an industrial revolution at its inception. But science occurs as part of culture, not distinct from it, and, as we have seen with the emerging issues of ownership, these new biotechnologies face not just scientific or practical problems, but the challenges of bringing them into society.

4. The Case for Progress

'It is easier to fixate on the threat than
embrace the opportunity.'

Rob Carlson, 2011

The threat of violence was not fulfilled. In fact, what was billed as
being an event marked by vandalism and direct action turned out
to be, by all accounts, something of a pleasant family day in the
country. On 27 May 2012, following a few weeks of publicity,
negotiation and public pleas, the clash between anti-genetically-
modified-foods activists and scientists happened on a sunny
Sunday in the green fields of Rothamsted, Hertfordshire, and featured ice creams, singing and some attempts at rousing rhetoric.
The activists, 'Take the Flour Back', had declared in their propa-
ganda that there isn't enough scientific data to support farming
GM crops, and that trials of genetically modified wheat in
Rothamsted's fields should be halted and indeed the crops
destroyed. They had declared that on this day they would 'decon-
taminate' the fields of GM wheat, as has happened on several
occasions in the last few years in the UK, meaning that they would
rip the experimental crops out and raze the fields. The scientists
leading the trials publicly pleaded and debated with them, but the
plan of direct action stayed in place.

The wheat in those fields carries several genetic modifications.
Some are for lab admin purposes (so that the functional experimental seeds can be selected from others). But the main man-made
gadget is the introduction of a genetic circuit that produces the
chemical E-β farnesene, or EβF (which, by pure coincidence, is

very closely related to the chemicals being made for synthetic diesel mentioned in Chapter 1). EβF acts as a klaxon for aphids, a pheromone that they pump out when being assaulted by predators such as ants. It tells other aphids to head for the hills. Aphids produce EβF themselves, but in a classy evolutionary move more than 400 plants that aphids like to occupy also produce it, as a sort of 'beware of the dog' sign on a canine-free house. Not only does it repel the aphids, it also prompts them to give birth to winged offspring, which can flee the perceived threat of danger. The experimental wheat has been created to produce EβF with the specific purpose of reducing the amount of aphid-slaughtering pesticide that farmers would normally need to use.

One earlier study had shown that EβF-engineered wheat had little or no effect on repelling aphids, but that was under lab conditions, which can be qualitatively different from plants in an open field, and the Rothamsted trial was designed as an attempt to address that exact question. A plaintive plea in an open letter by the scientists conducting the research pointed out that what Take the Flour Back had planned was 'reminiscent of clearing books from a library because you wish to stop other people finding out what they contain'.

Despite the decidedly undramatic unfolding of events in Rothamstead that day, Take the Flour Back were unrepentant, and issued a statement afterwards saying:

> We wanted to do the responsible thing and remove the threat of GM contamination; sadly it wasn't possible to do that effectively today. However, we stand arm in arm with farmers and growers from around the world, who are prepared to risk their freedom to stop the imposition of GM crops.

This small saga reflects the state of play in the public perception of modern biotechnology, and in many ways its entire forty-year history. Take the Flour Back are vocal and noisy, and generate plenty of publicity. But it is difficult to gauge how much they genuinely echo the public mood, or the numbers of the farmers

with whom they claim solidarity. In many ways, they draw inspiration from the Luddites. Though the name has evolved to refer to anyone who rejects technology, the first Luddites were artisan clothmakers who felt threatened by the advent of new mechanical looms. In the space of a few years at the beginning of the nineteenth century, they met by moonlight on the Yorkshire moors in northern England, broke into factories and smashed up the looms that had taken their jobs.*

For genetic modification, this kind of direct action is rare, but not unheard of: Take the Flour Back's website claims that 'Between 1999 and 2003 at least 91 GM trials were damaged or destroyed.' The sentiment behind this has roots as old as the technology itself, as we shall see below. In the UK, the potential introduction of genetically modified food in the 1980s was met with horror from campaigners, some of whom felt that to mess with nature in such a way was morally and intrinsically wrong. The press fuelled the revulsion by creating the phrase 'frankenfoods' – a peculiar portmanteau moniker, given that Dr Frankenstein was the creator of Mary Shelley's unnamed monster. But labels like that have power. Just as the 1972 thefts at the Watergate building in Washington DC somehow greenlit the press to forever suffix '-gate' onto any political scandal, in the last few years we have had headlines with frankenmice, frankenfish, frankenbugs and frankencrops in reference to anything genetically modified.

As synthetic biology grows scientifically, it is drawing attention from all quarters, and it has inherited the antagonists of generic genetic engineering. So in order to address the concerns and ire that synthetic biology raises, we need to look at the much broader picture of the public perception of biotechnology in general, and its short but world-changing history. We are beginning to see

---

* That particular movement was short-lived but weighty, and was crushed with ruthless force. Legislation was introduced in 1812 that specifically outlawed vandalism to machinery and made it a capital crime. Seventeen Luddites were hanged a year later, and many were sent to what was then the penal colony of Australia.

groups who have historically led the charge against GM, such as Action Group on Erosion, Technology and Concentration (ETC) and Friends of the Earth, taking very specific aim at synthetic biology and publishing pamphlets alleging that the products of synthetic biology are untested and poorly understood, and that they threaten all manner of human and ecological scenarios. Immediately following the publication of Craig Venter's Synthia (it was in fact ETC that bestowed *Mycoplasma mycoides JCVI-syn1.0* with that more friendly nickname), President Obama commissioned a report from synthetic biology experts to assess any potential threat, challenges and benefits from the whole field. The report, published in late 2010, addressed many of the issues by consulting with various publics and professional scientists, and drew up five categories by which to observe its progress: public beneficence, responsible stewardship, intellectual freedom and responsibility, democratic deliberation, and justice and fairness.

In assessing by these criteria, the commission decided that existing regulation was sufficient to allow synthetic biology to grow safely for the maximum benefit, whilst minimizing any potential threat. The principle of 'prudent vigilance' is cited, which roughly translates as 'let's keep a close eye on this'. But it also recommended openness and innovation through sharing, recognizing the open-access, patent-free ethos of BioBricks described in the previous chapter, and a coordinated approach across the whole field. It was a fillip to a whole field that acknowledged the potential of synthetic biology and strived to nurture growth.

But not everyone saw it like that. The commission drew considerable ire from campaigners. In response, fifty-six organizations led by ETC and Friends of the Earth immediately published an open letter to the presidential commission decrying the report for

> ignoring the precautionary principle, lacking adequate review of environmental risks, placing unwarranted faith in 'suicide genes'; and other technologies that provide no guarantee against the

escape of synthetic organisms into the environment, and relying on industry 'self regulation', which is the equivalent of no independent oversight.

Regulation for genetic modification is needed and does exist. In the USA, three federal agencies rule over the use of GM plants: the United States Department of Agriculture presides over the possibility of crops becoming weeds; the Food and Drug Administration on their entering and subverting the food chain; and the Environmental Protection Agency judges on plants with pesticide powers.

In 2012 the GM opponents increased their numbers to a coalition of 111 activist groups and put together their own report, which further denounced the field of synthetic biology. That number may sound like a lot, but many of these organizations are small groups of a handful of individuals, and so it is difficult to assess how representative their view is of the general population. Again with charged language and robust assertions, their conclusions were that more oversight and more regulation were needed, and that there should be a full suspension on the release and commercial use of synthetic cells.

It might be useful to look here to the origin of biotechnology to help understand the context in which this opposition is rooted. The epicentre of synthetic biology is at Stanford in California, where the BioBricks Foundation is headquartered, and in the mid 1970s it was also the birthplace of the founding technology. The discovery of restriction enzymes, those bacterial cut-and-paste tools with which this whole story began, had enabled an experiment by Paul Berg where he cut out parts of the genome of one virus and inserted them into another. This baby step into the world of genetic engineering earned him a Nobel Prize in 1980. But he was cautious about the implications of this new technology, and he held back from completing the experiment for fear of creating a monster, and putting his colleagues and the wider public at risk. Instead Berg called for a moratorium on this neonate field. In 1975,

he and others, at the behest of the US National Academy of Science, drew together a small international conference in Asilomar on California's Monterey Peninsula.

The Asilomar meeting is now regarded as a paragon of scientific responsibility when dealing with new and dual-use technologies, that is, new tools that have the potential to be used for both positive applications and by malefactors. The background culture in which it occurred featured the paranoid terror of blockbuster films such as *The Andromeda Strain* (1971), where an alien microbe caused madness and death indiscriminately and tapped into post-Watergate anxiety. Berg and others who were beginning to tinker with genetic code met with a sense of open caution in discussing where to go next. Journalists, lawyers and scientists also convened, and together they bashed out the arguments over several days. In the end, they drew up guidelines which suggested new research, open dialogues and drawing on expertise from experts beyond molecular biology, such as in infectious diseases and microbial ecology. They concluded that the 'new techniques, which permit combination of genetic information from very different organisms, place us in an area of biology with many unknowns'. That is a sentiment that applies as much today as it did then. In fact, the nature of science in general should foment by definition a culture filled with unknowns. But the sense of potential harm was judged to be outweighed by the potential benefits, and they recommended qualified progress:

> it was agreed that there are certain experiments in which the potential risks are of such a serious nature that they ought not to be done with presently available containment facilities. In the longer term, serious problems may arise in the large-scale application of this methodology in industry, medicine, and agriculture. But it was also recognized that future research and experience may show that many of the potential biohazards are less serious and/or less probable than we now suspect.

But the ideological rift between biotechnologists and their

opponents was formalized shortly afterwards. Just as in 2010, in 1978, not long after Asilomar, public opposition to the way genetic engineering was progressing began to emerge. There was also a falling-out of various scientists with environmental activists, including Friends of the Earth, with whom they had previously been allies. The terms were much the same as they are today: the degree of regulation required to enable safe progress in DNA-splicing research.

As the battle began, an anonymous scientist expressed the sentiment that 'some of the environmental lobbies are in business to peddle paranoia', and that their behaviour was merely thuggery. 'I fear,' James Watson told *Science* in 1978, 'that such groups thrive on bad news, and the more the public worries about the environment, the more likely we are to keep providing them with the funds that they need to keep their organizations growing.' Another scientist suggested, this time off the record, that 'these private agents of the public interest are not elected, nor are they necessarily in touch with the views of rank-and-file members of the groups they speak for'. It is quite possible that the same situation exists today, with popular support being garnered by shock tactics, publicity stunts and emotive assertions, rather than an attempt to engage in an honest and public dialogue, as we have seen in Rothamsted, and will see later in this chapter.

Paul Ehrlich, a prominent scientist and until that point member of Friends of the Earth, reiterated the view that 'the potential benefits from recombinant DNA research are so great that it would be foolhardy to restrict such research largely on the basis of imagined risks'. It is true that no successful acts of bioterrorism have occurred since Asilomar, and we will come on to that shortly, but as the industry of genetic engineering bloomed and mutated into synthetic biology, the potential problems remained the same, and access to the technology is considerably easier than in the mid 1970s. In 1975 you could fit every single expert on recombinant DNA into a decent-sized pub, let alone a conference centre. Today, the same technology has been democratized to the point

where everyone who has ever performed DNA manipulation would populate a small country. Romanticizing the past is probably not very helpful, but it seems that the climate in which the birth of molecular biology took place was significantly different from that today. Biotechnology was newborn then, and not the scientific and commercial behemoth that it is today. A US presidential report in April 2012 put the revenue of the 'bioeconomy' in the US in 2010 at $100 billion. It is only set to grow.

The actions and motivations of Take the Flour Back, and the policies of Friends of the Earth and ETC now, are not significantly different from those expressed forty years ago at the birth of genetic engineering. But the field has exploded. So the question is whether the original opponents were prescient in their opposition, or if their rhetoric is as irrelevant as it is unchanged, when the science has progressed. Here, we'll look at some of the major claims and scenarios.

Biological Weapons

The specific case of weaponizing engineered life forms is one of the key arguments put forward in opposition to both genetic engineering and synthetic biology. And our advancing skills in DNA technology have kept the fear of bioterrorism very much alive. Weaponizing life is not a new phenomenon. The use of living things (other than humans) to create havoc has a rich but utterly ignoble tradition that predates our conquest of DNA by millennia. Hannibal used his attack elephants in Roman antiquity. A Mongolian chieftain invading India in the fourteenth century dispatched camels ablaze to charge scimitar-wielding elephants. Even in the modern era, rats, cats, pigeons and dogs have all been used as either bomb-detecting or bomb-bearing devices. In the Second World War, the Russians had some arguable success by training dogs to run under German panzers while carrying an explosive triggered by a knock to the tanks' underbelly.

Nowadays, the potential threat of living weapons is no less bizarre, but far more potent, and, as with all aspects of genetic engineering, the key difference is the introduction of precision control of the language of life. Much like most biology, bioterror has been reduced in size from the use of animals to the microcosm of genetics.

The universality of genetic code is something that viruses have exploited throughout life's history, as they insert their own DNA into an unwitting genome to usurp that host cell's own biological mechanics. We live in a world that has always been plagued by disease. Death by infectious agents far outnumbers death at the hands of another human, and our mastery of genetic manipulation has presented the possibility of terrorism with these tools. In 2006, a reporter from the *Guardian* newspaper in the UK attempted to make a point by acquiring parts of the smallpox genome. This is a disease that has been eradicated from Earth.* Its genome sequence is freely available online, as all publicly funded genome sequences are, and the intention was to question how easy it would be to build a maleficent creation. By ordering a few sections of the genome of *Variola major*, the virus that causes smallpox, it was suggested that a whole genome could be stitched together from smaller pieces. The short, seventy-five-base sections were manufactured by an unremarkable UK gene synthesis company and sent to a residential address in north London, unchecked for their weapon potential. As responsible investigative journalists, they introduced deliberate errors, and the sequences ordered encoded part of the virus that was itself not toxic. Nevertheless, the story ran with startling revelations:

> DNA sequences from some of the most deadly pathogens known to man can be bought over the internet, the *Guardian* has discovered. In an investigation which shows the ease with which

---

* Two smallpox samples are kept in cold storage for scientific purposes; there is an ongoing discussion about whether these should be kept or destroyed, primarily because of the risk of bioterror that they embody.

terrorist organisations could obtain the basic ingredients of biological weapons, this newspaper obtained a short sequence of
smallpox DNA.

Shocking though this might initially sound, it was a ham-fisted
stunt at best. Its only real triumph was to show that some DNA
synthesis companies were not particularly vigilant in taking orders
from members of the public. And their claim to have acted respon-
sibly by building errors into the code was lip service. Once you'd
got hold of the sequence, correcting the errors would be a trivial
matter for any competent genetics graduate student.

But the overall premise was shaky. Using this technique of
stitching together seventy five-letter fragments to make up an
186,000-letter genome is a tall order, akin to reassembling a shred-
ded document twice the size of this book. 'Theoretically possible',
the article asserts, but in reality paralysingly arduous. In 2002, a
New-York-based team did successfully assemble a working polio
virus from mail-order synthesized segments of DNA. But that
genome is only 7,500 bases long and still took a team of molecular
biology experts two years. These are not directly scalable problems. Recall that Venter's Synthia in 2010, with its 582,000-letter
genome, took twenty people ten years at an estimated cost of
$40 million. There is a wide range of actuality when discussing the
'theoretically possible'.

Anti-synthetic biology activists ETC produced a report on synthetic biology in 2007 that refers to the *Guardian*'s stunt, and
manages to enhance the theoretical possibility of danger further: 'A
commercial outfit could theoretically crank out the entire DNA
for a synthetic version of *Variola major* in less than two weeks, for
about the price of a high-end sports car.' This might be 'theoretically' possible, but only if its meaning is stretched a long way
beyond the practically feasible. As synthetic biology expert Rob
Carlson says, 'It is easier to fixate on the threat than embrace the
opportunity.'

It is important to acknowledge that there is a potential threat,

and that synthetic biology and genetic engineering are dual-use technologies. Diseases consigned to history *can* now be reconstructed in the lab, and ones that we haven't conquered yet can be engineered to be more dangerous. We are capable of eradicating diseases, such as smallpox, thanks to vaccination. Polio is likely to be next and, in time, diseases that kill us today will only be of historical interest for our children. Though it is unequivocal that science and medicine have transformed humanity's survival, it is not news to point out that millions die each year from diseases that have no cure. But how realistic is the possibility of genetic engineering being the tool that malefactors use to bring the terror? The *Guardian*'s smallpox story ran six years ago and, as we have seen, the technology in question is evolving at a breathtaking rate. In 2012, the game changed yet again.

Killer Flu

Influenza comes primarily from birds. The virus annexes a host cell's workings to enact its own genetic program and make its own proteins. In flu viruses, two of these proteins sit on their surface as they drift from victim to victim: haemaggluttinin (H), which hooks onto a target cell to gain entry, and neuroaminidase (N), which newly made virus particles use to get out again. Variations in this pair of break-in and bust-out proteins gives the various flu strains their names, such as H5N4. For the most part, H5N4 flu remains a bird disease. That's not to say that it doesn't cause symptoms, it's just that humans won't make new viruses, so we can't spread it. But it is constantly evolving, and finding new ways to spread. Once in a while, a strain will make the evolutionary leap to become a human disease, with both symptoms and the ability to transmit from person to person, with sometimes apocalyptic results. In 1918, the flu pandemic caused by an H1N1 strain infected billions, and caused the deaths of fifty million people.

In 1997, patients in Hong Kong started dying from a new strain

that hadn't been found in humans before: H5N1. This was a version that abounded in Chinese open-air chicken markets, to the extent that it was considered endemic. The very real worry was that, as it was a new strain, humans wouldn't have adapted any immunity to it. Flu's accelerated evolution can happen when two different strains infect the same cell, where they can trade genes. So if H5N1 swapped genes with a strain that struck humans and was transmissible from person to person, we would be looking at the beginning of disaster.

Knowing full well that highly infectious chimera strains are a theoretical possibility, teams of flu researchers set about fulfilling Feynman's maxim: 'What I cannot create, I do not understand.' Scientists in the Netherlands and North America began pre-empting nature by attempting to assemble the new combinations exactly as they might evolve to brandish pandemic-causing qualities. Yoshihiro Kawaoka from the University of Wisconsin-Madison and Ron Fouchier from Erasmus Medical Centre in Rotterdam each designed flu viruses that would infect mammals, in this case ferrets. Kawaoka's design shuffled two decks of viral genes to put together one that not only would break into cells in a ferret's nasal passages, but could replicate and infect other ferrets via airborne droplets. Fouchier's design was similar, but had just five genetic alterations, all seen in other natural strains, that enabled the virus to clamp onto ferret cells and infect. These also were passed on to other ferrets via a sneeze. Neither strain was lethal, though Fouchier's creation caused death when delivered in concentration via an inhaler. Given the importance of this work, these two studies were sent to the two most important scientific journals in the world, Kawaoka's to *Nature*, and Fouchier's to *Science*. What followed was a tortuous tale of negotiating a delicate path through frontier territory.

The US National Science Advisory Board for Biosecurity (NSABB) has oversight on exactly these types of scenario, and, although they have no power to compel, in December 2011, they recommended that the two papers should be published but

censored to 'not include the methodological and other details that could enable replication of the experiments by those who would seek to do harm'. NSABB are a collection of scientists, lawyers and policymakers who advise on dual-use research in biology. Their initial judgement was that both papers could be published but with methods and the genetic sequences themselves redacted. This prompted some serious and necessary hand-wringing by all parties involved. Mostly, when controversial science is reported, the debates tend to be between researchers and non-scientists. But this time round, unusually, there was intemperate disagreement between scientists on whether to publish in full, redacted or not at all. As had happened at Asilomar, this confrontation prompted a dialogue open to the public. And also emulating the actions of Paul Berg in 1975, Fouchier and Kawaoka both called for a sixty-day voluntary moratorium on flu research in order to work out what the right thing to do was. The World Health Organization stepped in, and after much gnashing of teeth the NSABB changed their minds and recommended full publication of the Kawaoka paper and a clarifying edit of Fouchier's. By this time, *Nature* had reached that decision independently.

These studies show the very real possibilities of new and dangerous flu viruses emerging in the real world. They also show that to weaponize flu virus would be a feat of genetic engineering, not something that anyone could do with any degree of ease. Those studies took major investment and tens of thousands of highly skilled man-hours. For a potential terrorist, this was not the only disadvantage. A transmissible mutant mammalian flu virus could not be targeted at any one group of people, so any attempt to murder a specific population would be impossible.

Nor do viruses respect geography or nationalities. One theory behind the global spread of the devastating 1918 flu was that it began at a poultry farm in Kansas that supplied a local military base with chickens, and that the soldiers then took it with them as they were sent off to fight the first global war. These days, with world travel and a virus emerging in a dense conurbation such as Hong Kong, the

potential for reaching humans all over the world is even greater. There is a threat from accidental release of transmissible viruses from a lab, or malevolent intent. Even if you were hell-bent on creating a synthetic indiscriminate killer virus, rather than the millions it would take to create, the very well-proven process of artificial selection would be much more efficient, and your starting place would be a poultry farm with dubious hygiene. Poor chicken-farming practices are far more likely breeding grounds for killer flus.

The utility of flu as a weapon is also questionable. If mass murder is your goal, then flying a passenger plane into a skyscraper is considerably easier. But the whole point of synthetic biology is to reduce the entry level for artisan genetic engineering, and the open culture of BioBricks has meant that gene manipulation that was impossible a decade ago is now school science. Does this make the entry level for bioterror lower? Currently, there are no bricks in the library that could be directly associated with a biological weapon. But it is possible to build a nail-bomb out of components that are individually innocuous enough and can be bought from a standard hardware shop. Indeed, one can make all kinds of explosives using simple household items, if one were so inclined. ETC's 2007 report on synthetic biology describes it as 'genetic engineering on steroids' and that ultimately it will mean 'cheaper and widely accessible tools to build bioweapons, virulent pathogens and artificial organisms that could pose grave threats to people and the planet'. Maybe there is a flicker of truth in that sentiment, in that any progress in synthetic biology will allow the development of new tech with dual use potential. But currently, and indeed for the foreseeable future, the reality of a genetically engineered weapon relies on the most generous use of the words 'theoretically possible'.

This tale of the killer flu experiments also raises a very fundamental question about the nature of science. One of the key principles in scientific research is the freedom to research any subject and for the results to be published. Perhaps with the advent of such potent dual-use technologies, this era is drawing to a close.

But to censor scientific information about potential agents of terror is to ignore two problems. The first is that, by studying how pathogens work, we understand better how to deal with them. This might be in the face of a terrorist threat, or simply to help treat patients, especially when virulence harbours epidemic potential. This needs to happen in an open, unrestricted way, as the richest, most productive and creative science comes when information is unfettered by barriers and shared freely. In doing just that we equip ourselves to tackle exactly the same threat should it come at all. Knowledge of how new virulent or even weaponized life forms work will ready us for both natural and created threats. On publication of the Kawaoka paper, an editorial in *Nature* ruled that 'A paper that omits key results or methods disables subsequent research and peer review.' The final point is also a practical one. Keeping the information under wraps and for your eyes only is simply not enforceable. *Nature* was aware that there were many versions of the paper circulating outside of the official channels of the publication process, which is more and more common in the Internet era. The editorial went on to say that it could not 'imagine any mechanism or criterion by which to sensibly judge who should or should not be allowed to see the work'.

In a sense, these two studies were the first and most obvious experiments on a flu threat. Evolution is tenacious, and life forms strive for immortality. In terms of understanding how that will unfold, we are always playing catch-up. We cannot predict every possible natural threat, but we should do what we can to arm ourselves against the ones we can anticipate. Any knowledge in this arms race is better than none. That, to my mind, is good science, and good policy. It is using the precautionary principle in an aggressive way. It was unclear until these two flu studies whether it was even possible for these strains to jump the species barrier. They also provide clear genetic flags for public health sentries to look out for in emerging flu strains. By working out how to upgrade a flu into one that can cause a pandemic, we arm ourselves against the very real possibility of it happening naturally.

Agenda Politics

Another line of attack is that genetically modified food might be bad for us to eat. In biology we look for mechanisms that explain observations, and it's not easy to imagine what mechanism would result in an adverse effect on our health after eating a modified food. Nevertheless, if it were an observable phenomenon, it would warrant explanation. Opponents of GM often cite the idea that eating GM foods might cause us harm, even though it has been thoroughly refuted in many independent reports. And with many modified foods having been approved a long time ago, there is also more evidence than a scientist could wish for demonstrating that they are safe to eat. According to one estimate, two trillion meals containing genetically modified food have been consumed in the two years up until 2011. The US Department of Agriculture reckons that GM revenues in 2010 were $76 billion, with the products being part of the human food chain.

Despite this, the arguments rumble on. In September 2012, a small study published in a minor journal hit the headlines and stirred the GM debate yet again. A French team of scientists, led by a molecular biologist called Gilles-Eric Séralini, fed a type of genetically modified maize to rats over their whole lifetime, and observed any effects on their health. The modified feed had been developed by biotech company Monsanto to resist a widely used herbicide. It has been widely approved as food in humans and animals around the world, and previous studies have shown no adverse health effects with this particular modified maize (called NK603, previously Roundup). Indeed, in 2012, what is considered to be the gold standard in these types of trial, a systematic review of multiple long-term studies, also concluded that there were no adverse health effects for a diet of GM food. In Séralini's study and the surrounding press, they claimed that some of these previous studies were qualitatively different from theirs. Their paper graphically showed that the rats used were profoundly affected by their

GM diet, suffering grotesque tumours and dying significantly before the control rats.

Immediately, though, the study began to unravel. The authors and the paper were bluntly criticized for the quality of the work, and also the way it was reported. With a swiftness enabled by the Internet, scientists began condemning the data, saying that the experimental design, the statistical analysis and the way the data was presented were all substandard, and some expressed surprise that such a paper was published at all. Specifically, critics noted that Séralini had used rats that were already prone to developing tumours, as well as inadequate numbers of control rats compared to experimental ones.

It soon emerged that the release of the results was part of an orchestrated PR campaign that included a book by Séralini, who has a track record of anti-biotech activism, and a television documentary. The paper itself was not freely distributed to the press before publication, which is extremely unusual, and the type of thing that makes science reporters highly suspicious: most will seek comment from other scientists, under embargo. With Séralini's paper, a signed agreement was required for those who were offered access to the paper, which came with another highly unusual stern warning: 'A refund of the cost of the study of several million Euros would be considered damages if the premature disclosure questioned the release of the study.'

An organization called the Sustainable Food Trust had spearheaded a campaign to spread the work as far and as wide as possible, including a dedicated web page resource, with their motivation being the promotion of what they call 'Good Science'. They encouraged supporters to pass on prefabricated messages of support on the influential social media site Twitter. Almost all of the mainstream press had covered the paper, although, interestingly, some of the coverage, such as by the BBC, had included some of the immediate profound criticisms from scientists. A few days later, the European Food Safety Authority issued a statement: 'The design, reporting and analysis of the study, as outlined in the

paper, are inadequate,' adding that the paper is 'of insufficient scientific quality to be considered as valid for risk assessment'.

But the PR offensive had done its work. Many press outlets published the story uncritically. The French news weekly *Le Nouvel Observateur* ran a cover which shouted out 'Oui, les OGM sont des poisons!' ('Yes, GMOs are poisonous!'). The French prime minister declared that France would defend banning these genetically modified crops across Europe (though he had the good grace to specifically add the caveat of needing to validate the results). Bearing in mind the distinctly unusual way the paper was published, the results are not invalid until other studies show them to be. That is not to say that they are correct, and, given the strong criticism of the methodology itself, it seems fair to say that, at the very least, Séralini's study demands the most rigorous replication of the results, releasing the original data for independent analysis, and sureing up the methodology to establish beyond doubt its validity.★

Within minutes of publication, many were forensically investigating not just the paper, but the way it was published. To get traction for an idea is the aim of such a PR exercise, and, in this case, they succeeded partially, in the face of the rapid scrutiny of critics. Some have suggested that this case study marks a turning point in the conflict over GM, as the condemnation was so quick and so severe that some of the mainstream press included it. But

---

★ To get published in a scientific journal requires 'peer review', where other expert scientists rate the validity of the proposed publication, and recommend for publication, or otherwise. But it is not a hallmark of scientific truth, only a recognition that the experiment is worthy of entering the academic literature. In this case, some have disputed whether Séralini's paper should have been published at all. Historically, critical feedback came via the cumbersome method of publishing correspondence or replicating experiments in academic journals. But now there is a form of post-publication peer review enabled by the Internet, where papers are scrutinized in blogs and columns and discussed openly. It's not formal, and is subject to the wildness of a frontier without gatekeepers, but it is quick, and scrutiny was precise and harsh.

just like rumours, once scientific ideas like this are out in the open, they are very difficult to correct, regardless of poor methodology, or knowing the agenda of the perpetrators, or anything else.

There are volumes and volumes of publications on GM and in these pages I am mentioning the tiniest handful from the very recent past. This is not to cherry-pick the data to fit a particular stance, a crime of agenda which I denounce, but to illustrate how research into GM research is subject to manipulation, and thrust into the limelight because it evokes passionate and polarized responses. Public and political opinion can and has been skewed by the way these types of campaigns are reported, and the deliberate polarization of public debates detracts from nuanced, informed and necessary analysis of the application of biotechnology.

Malaria

To focus on the possibility of harm from GM foods or bioterror is to fixate on a threat, whether looming or currently implausible. As an industry instead providing opportunities, the nascent field of synthetic biology has so far yielded few genuine success stories. In this book, I have expressed enthusiasm for its potential, but in this chapter and elsewhere I have looked at some of the harder realities of applying groundbreaking science into society, which is explicit in synthetic biology's mandate. There have been innumerable examples of the scientific successes of genetic engineering in its broadest sense, not least in terms of understanding the basics of biology and the underlying causes of thousands of diseases. The work I once did in the lab helped to identify a form of blindness in children, and this would not have been possible without the genetic modification of mice.

Synthetic biology is yet to deliver on those scales. But it is set to start, and just like any discipline that involves extremely complex and constantly changing science is subject to the scrutiny of not just people and politics, but also market forces and economics. The

story of the synthetic biology company Amyris and their stalling attempt to turn synthetic bio-diesel to market is described on page 24. But probably the single biggest success story of synthetic biology, along with the quagmire of real markets, comes from exactly the same team, labs and genetic circuit board.

Over the spread of human history, malaria has been a constant lethal presence. Some estimates put the number of people who have perished at the hands of plasmodium, the malaria parasite ferried by mosquitoes, in the tens of billions, depending on how you define human species. Today, a quarter of a billion people are infected every year, and the highest estimates are that more than a million will die, mostly young children. Apart from the staggering human cost, the World Health Organisation (WHO) estimates that the cost of malaria to GDP for sub-Saharan Africa is in the order of $100 billion dollars since the 1960s. The impetus to reduce the burden of malaria is profound.

Since the seventeenth century, quinine, extracted from South American cinchona trees, was the preferred treatment, though it came with a suite of unpleasant side effects. A chemical cousin called chloroquine usurped quinine as the standard anti-malarial drug following the Second World War. But, as with all life forms, the parasite wants to continue to live, and it does so by evolving. Mass treatment with chloroquine meant that strains of plasmodium that were resistant to the deathly chloroquine emerged in the 1950s, and spread across the world in a bid to ensure its own survival in the face of man-made extinction.

The current drug of choice in malarial areas is a small molecule called artemisinin. It is extracted from sweet wormwood, an Asian camphorous shrub, which is grown all around the world and has been used in folk medicine for centuries. As a treatment for malaria, artemisinin is fast and effective. WHO specifically prohibit its use on its own, for fear of unintentionally nurturing and selecting strains of plasmodium that are naturally resistant. Instead they recommend that artemisinin forms the key player in combined therapies.

Getting enough artemisinin is a pricey and finicky business. It has to be farmed quickly and in very specific ways. From an economic point of view, the very fact that it is cultivated means that the fields of wormwood are competing with crops for food or feedstock.

Drugs are created in chemistry labs. You start with basic chemical ingredients, which can be bought from chemical companies or harvested from the living world. The production of drugs is like forensically accurate cooking, as each ingredient is mixed in to systematically add or subtract bits of molecules until you have the drug desired. Artemisinin is not a large molecule, but unfortunately it's also not easy – or more importantly cheap – to synthesize using straightforward chemistry. Early in the process of working on synthetic circuits to create biosynthesized diesel, Stanford University researcher Jay Keasling was alerted by a student to the fact that one of the steps in their chemical pathway was also a key link in the chain that could produce synthetic artemisinin. While he tried to build a genetic circuit that would create diesel, his team also pursued another that would construct artimisinin. They published the first successful circuit for artemisinin synthesis in yeast in 2006, having switched from bacteria in 2003. It's built from twelve genes taken from three different organisms.

In both cases, upscaling to industrial production was inherent in Keasling's aims, again marking the intent not merely to create, but to apply, scientific research to real-world scenarios directly. It's a strange thing to see in a scientific paper, but from the start reducing the cost of production was clearly set out as a mission: 'Industrial scale-up will be required to raise artemisinic acid production to a level high enough to reduce artemisinin combination therapies to significantly below their current prices.' This exemplifies the problem-solving, engineering sensibility built into synthetic biology. It denotes the application as the goal, not merely to produce a functional drug, but also to do it cheaply.

Keasling's rationale for this apparent altruism stems from the market forces that had confused artemisinin farming in the

preceding years. At the end of the twentieth century, there was a great paucity of wormwood farming, and so the market drove the price of artemisinin up. Then, spotting a gap in the market, thousands of African and Asian farmers started growing wormwood. And the price dropped. But even with doses being artificially maintained by government clinics at $1 a go, more than half of patients would buy more expensive versions out of convenience from market stalls not acknowledging the government intervention. Unreliable supply, fluctuations in prices and variable yields drove the prices back up, and this pattern looked set to continue.

Keasling set up Amyris to develop artemisinin production, along with its diesel sibling (which has tanked, at least temporarily). Following on from that, the Bill and Melinda Gates Foundation, the charity set up by the Microsoft billionaire and his wife, chipped in with $46 million to the charity Institute for One World Health, which worked with Amyris to realize industrial-scale production. Amyris has run this project on a not-for-profit basis and joined up with organizations such as the Global Fund to reduce the cost and distribute as much artemisinin (as part of a combination therapy) as possible, as cheaply as possible. What that means is driving the cost down to less than fifty cents per treatment. Amyris granted the pharmaceutical giant Sanofi-Aventis a royalty-free licence to produce vats of synthetic artemisinin, which are expected to hit the shelves in the next two years.

Implementation of a cocktail of anti-malarial drugs is the WHO's strategy for combating malaria. This way, resistance doesn't evolve against artemisinin and declaw its therapeutic power, as it has for its ancestors. But companies disregard the WHO's strategy. Since the constituents of combined therapies can be sold individually, there is profit to be made in ignoring their advice. In 2009 the WHO felt compelled to issue a briefing that warned companies to stop marketing artemisinin as a single therapy. According to the WHO report, the motivation is financial, as therapies can be sold individually, ignoring the

recommendations. Andrea Bosman, part of their Global Malaria Project, told *Nature*: 'It's terrible. Who says there is no profit to be made in malaria? When you see the number of companies operating in Africa, and the diversity of products, you'd just be amazed.'

The reality of applied science in the market provides the major caveat to this story. Exactly what WHO scientists were trying to avoid came to pass. The first cases of artemisinin resistance had been discovered in Myanmar in 2004, just as had emerged to chloroquine in the twentieth century. Containment of the evolved parasite became the priority, but by 2012 patients with limited responses to artemisinin were recorded in Cambodia, indicating that the resistance had spread (or possibly evolved separately). With luck, the principles of component switching and circuit tweaking that characterize synthetic biology will make the road to modifying malaria treatment easier and smarter in the years to come.

In the immature artemisinin farming market, prices have fluctuated wildly in cycles of boom and bust. Though this instability is undesirable, part of the argument against the establishment of a synthetic production base is that it would result in the displacement of thousands of wormwood farmers, as their livelihood is ruined by a greatly reduced cost of synthetic production. This is one of the attacks that anti-synthetic biology campaigning groups put forward, and may have some legitimacy. In a 2012 briefing pamphlet specifically criticizing the synthetic artemisinin project, ETC asserted that artimisinin demand could be met solely with an increase in wormwood cultivation, and therefore a synthetic production was not only unnecessary, but will take production control from small-holding farmers and hand it to large Western pharmaceutical corporations.

Part of this argument runs the risk of falling foul of what's known as the 'Luddite Fallacy', the idea that technological unemployment follows technological advances, as first mechanization and then automation replace workers. It's a fallacy, at least in

the long term, because, if it were true, as economist Alex Tabarrok points out, 'we would all be out of work because productivity has been increasing for two centuries'. At the same time, the combination of a not-for-profit enterprise to produce an affordable drug to treat millions of people must outweigh those concerns. The collaboration between Amyris, One World and pharmaceutical giant Sanofi-Aventis bears the hallmarks of corporate responsibility. They have agreed to work on a not-for-profit basis, and to limit supply to 50 per cent of the full artemisinin market, which leaves some space for traditional farming. Whether this collaboration will be the benchmark of humanitarian aid enabled by synthetic biology is yet to be seen, as, at the time of writing, synthetic artemisinin has not hit the markets. If and when it does, artemisinin will be the first genuine commercial product of synthetic biology, and, if successful, it will be puzzled over, studied and followed for years.

Should or Could? The Case for Progress

Science is primarily a public endeavour, not least because it is largely funded by the public purse, but also because the benefits of non-commercial scientific research are for all. As synthetic biology emerges, it will face all of the concerns that GM has had over the short history of biotechnology, and more as it grows. One thing seems to be essential: the discussions about synthetic biology and genetic modification must happen in public, and with the public.

The calls from the opponents of genetic modification and now synthetic biology for bans are unrealistic and destructive. They are designed to foment fear derived from an ideological position, and to enrage rather than engage. The eighteenth-century satirist Jonathan Swift suggested that it was not possible to reason someone out of a position that they didn't reason themselves into. If resistance to new forms of biotechnology are ideological, then the challenges of contesting those views with evidence are hard. There

are legitimate and serious concerns in any of the technologies described in these pages, but they demand rational, open and informed discussions.

GM crops are in the wild and they are in our food. Another significant challenge from green campaigners is that, once out in fields, GM crops can cross-breed with traditionally manipulated crops and the unnatural genes can out-compete their more natural counterparts in the open market of nature. This is a more legitimate concern, as we know that this can happen. As with the EβF wheat in Rothamsted, gene engineering in crops is frequently aimed at reducing the need for pesticides, by getting the plant to produce its own. Similarly, engineering a crop to be resistant to a weedkiller is potentially useful, as it means that farmers can spray their fields with a herbicide knowing that only unwanted plants will die. But so far, field experiments have produced mixed results. Some GM crops do appear to have the effect of reducing local biodiversity, with fewer wild plants and fewer insects to pollinate them. But it is not universal. In one experiment, GM maize appeared to increase biodiversity, whereas beet and rape did the opposite. That sort of result is not uncommon: biology is messy. But this should lead us to conduct *more* experiments, not to shut down or even vandalize them.

More and more scientists and politicians are now moving towards the stance that, in order to address mammoth population growth, poverty and incoming climate change, genetically modified crops will be utterly necessary. John Beddington, the UK Government's chief scientific adviser, told the BBC in 2011 that

> If there are genetically modified organisms that actually solve problems that we can't solve in other ways, and are shown to be safe from a human health point of view, and safe from an environmental point of view. . . then we should use them.

The precautionary principle is carefully threaded into that statement, suggesting the necessity of GM rather than a whimsical or commercial desire to introduce these creations. There is a simple

need underlying Beddington's comment: in changing environments, many of which are going to make land in poor nations less amenable to arable, we need new ways for crops to be hardier in difficult ground. Breeding is slow and cumbersome, and therefore is not a realistic option.

For synthetic biology, its potential for use in the service of humankind and the planet puts it in a similar camp. We don't know the full effects of introducing its products out of the lab and into the wild. One of the main demands of synthetic biology antagonists is that these cells and unnatural life forms should not be released. We have seen with GM crops the spread of their modified genes beyond the original host, though whether these have a detrimental effect is unknown. Could any of the products of synthetic biology cause havoc? The cancer assassin circuit described in Chapter 3 integrates into the genome of a virus, and delivers its lethal message after infecting a cancerous cell. It is designed to target only one specific type of malignant cell, and performs a calculation to precisely determine its cancerous nature. If it were capable of going native and joining the natural ecosystem, is there a risk of it entering other cells and destroying them? It is unimaginable that it could, because of the very precise nature of the design of the circuit, but theoretically not impossible, just tremendously improbable. We do know that genes swap in and out from microorganisms and viruses, and this can't be ruled out.

Could Synthia live beyond the lab? *Mycoplasma mycoides* is a natural pathogen, albeit it a minor, non-lethal one for goats. It might be able to survive outside of the lab, though Venter's team specifically wrote into its genome a chunk of code that rendered it incapable of infection, thus denying it its favoured habitat. But the remarkable tenacity of bacteria might suggest that it could gain functions upon exposure to other bacteria, as they swap their genes.

Could Amyris' artemisinin- or biofuel-producing yeast cells escape from their vats and wreak havoc on ecosystems? It is

unlikely, as they too are optimized for purpose, not for survival. They are not in the wild, or in field tests, though their non-living products will be.

All of these scenarios are not airtight, but after careful consideration the risks seem almost trivial, and just like with the outstanding questions in genetically modified farming, they suggest the need for more research, not less.

Knowledge is Value Free

In 2007, an editorial in the journal *Nature* made this comment: 'Many a technology has at some time or another been deemed an affront to God, but perhaps none invites the accusation as directly as synthetic biology. For the first time, God has competition.' It was written in broad support of synthetic biology, but carries a theological misunderstanding. Since Eve took a bite of an apple, God, if one believes in such things, has had competition. Throughout our existence we have challenged, manipulated and forged nature to our own ends. Our influence has shaped and defined this living world, hitherto governed by Darwinian rules. With synthetic biology we have the opportunity not to supplant those rules, but to create new lives specifically for purpose. That is not a call to defeat nature, nor to trample it underfoot any more than we have already done. In less than a fraction of a heartbeat of this planet's existence we have done more than enough to jeopardize our continued existence here. The living planet will go on cycling according to the fixed law of gravity, with or without us, its most creative and destructive offspring. But with smart, informed scientific decisions we have the ability to fix our past errors.

In 2005 on the thirtieth anniversary of the Asilomar meeting, Paul Berg commented: 'First and foremost, we gained the public's trust, for it was the very scientists who were most involved in the work and had every incentive to be left free to pursue their dream

that called attention to the risks inherent in the experiments they were doing.' One tenth of the conference attendees were journalists, who were free to observe the sometimes bitter altercations and fights between scientists. It was the scientists who raised the concerns and, by arguing them out in public, they ensured that the outcomes were cautious, progressive and uncontroversial.

It may be that current regulations will not be sufficient in the future, but they should be addressed then, not pre-emptively in a way that could prohibit progress. If 'prudent vigilance' is built into the structure of research funding, public engagement and the way the research is done and applied, then the precautionary principle is foremost in the way synthetic biology proceeds. Whether this works remains to be seen, but to halt progress is to deny the potential benefit that new technologies can provide. Furthermore, increased regulation, restricted progress and heavy monitoring drives actual research out of the hands of publicly funded scientists, because all of these things are expensive. Those levels of bureaucracy are handled comfortably by companies with commercial interests, who are not bound to have open dialogues with a public who might otherwise fund them.

Synthetic biology is moving at such a pace that many scientists are bewildered by its progress. To bring the public and politicians on board is a difficult task. But it is essential, as it is society which makes the decision about what we should and shouldn't do. 'Democratic deliberation' is the phrase used in President Obama's 2010 bioeconomy report. To this end, in the UK in 2010, major scientific funding agencies commissioned extensive research on what people know and think about synthetic biology. Key questions emerged:

What is the purpose?

Why do you want to do it?

What are you going to gain from it?

What else is it going to do?

How do you know you are right?

These are precisely the right questions that any scientist should ask of any project, so in that sense this type of dialogue is working.

But in reflecting on the triumph of the Asilomar meeting, Paul Berg also warned that an attempt to repeat past successes would be a futile gesture: 'By contrast, the issues that challenge us today are qualitatively different. They are often beset with economic self-interest and increasingly by nearly irreconcilable ethical and religious conflicts and challenges to deeply held social values. An Asilomar-type conference trying to contend with such contentious views is, I believe, doomed to acrimony and policy stagnation, neither of which advances the cause of finding a solution.'

I don't share this pessimism. I believe that scientific research should occur in the full glare of public scrutiny and that scientists should engage with publics of all levels of expertise. This way, with data in the open, and informed public conversations about the potential benefit and the potential harm that new technologies create, we foster a society where rational approaches to global and local problems are normalized.

There is a revolution happening, and it is in our hands. Synthetic biology promises benefits for all humankind's residency on Earth and on worlds not yet explored, far too great to ignore, repress or censor. There is youthful optimism, a remix culture of boundless creativity that believes these new technologies will help rectify the problems we face and, at least in some quarters, an unprecedented willingness to push that agenda.

We will face new and unforeseen challenges, some that we have created or at least nurtured. But we should strive to invent technologies that are not in conflict with nature, nor subvert or exploit it, but work alongside our sophisticated living world with its four-billion-year-evolved history. Humankind's progress is born out of our attempts to explore and understand a continuously changing landscape, and live within those means. Our exploration of the workings of living things over two or three centuries has given us the power to do things that could never have happened before. We

are engineering life forms to build fuels, drugs, treatments, the tools to explore our universe and boundless new living creations that can help our world and our tenure on it. Our responsibility is not to curtail that knowledge but to use it to better ourselves, and our living planet. Sometimes this gets framed as a question of whether we should do the things that we are capable of. The answer transcends either option: we must.

Afterword

New Languages

Language evolves. Samuel Johnson noted in his famous dictionary that all languages 'have a natural tendency to degeneration', though throughout history terminal decline has been somehow miraculously avoided. That anxiety about the way languages change – along with a suggestion that languages once, at some point in the past, existed in perfect and inviolate form – has been expressed in every generation throughout history by brabbling malagrugs, but really only reflects the fact that words and their meanings constantly change over time. Cromulent words are added, mangled, misappropriated and invented. Very occasionally, languages acquire new letters too.*

Evolution settled on an alphabet of just four letters and twenty words, and just a single language. This alphabet and lexicon† has remained static since the beginning of life on earth, a stable mechanism that is flexible enough to encourage evolution in a species but shields the individual from catastrophic changes to their genomes.

With the genetic code being the only language of life we know of, there is no possibility of importing letters or words from other

---

* Almost always, this pilfering is to cope with words on loan from foreign languages. French has virtually no uses for the letter *w* except in borrowed words, such as 'le weekend'. Welsh acquired *j* to cope with English and Norman words and names. Languages and their alphabets contract and expand like living things.

† A, G, C and T (U in RNA) arranged in genes into triplets, which together encode one of just twenty amino acids and three stop codons.

life forms; there is no direct analogy for the kind of borrowing and exchange we see between languages. However, genes and DNA – the words in our analogy, as opposed to the letters of the genetic code – can swap between species, particularly single-celled species of bacteria and archaea. But they can also be transported harmlessly into more complex creatures when viruses integrate their genomes into a host. It is estimated that up to 8 per cent of your DNA was once in the genome of a virus.

Just as we have difficulty in pinpointing the precise moment when a species comes into being, there was no point where we can say that English was born, nor any of the myriad languages or dialects that are and have been spoken in human history. However, *Kelkaj lingvoj ne evoluas; kelkaj simple elpensiĝas*, as Esperanto-speakers might say: some languages did not evolve; they were simply invented. Esperanto was a noble attempt to induce world peace by making a universal language, and survives a century after its creation with a few thousand speakers. In fact something like 900 languages have been invented from scratch, including Klingon, as spoken by the angry fictional warriors of the *Star Trek* franchise and a few hundred devoted followers.

Our command of the language of DNA and its mechanics has brought us to the point where we no longer need be content with the four-billion-year-old alphabet, lexicon and language of life. In the era of synthetic biology, we have begun the process of inventing new ones.

If you've ever had a cold sore, you've probably already used artificial DNA letters. The universal letters, nucleobases, are of course merely chemicals. A for Adenine is a collection of fifteen atoms: five carbon atoms, five nitrogens and five hydrogens. They're arranged into a hexagon attached to a pentagon with a few molecular bobbles protruding. There is nothing inherently special about that arrangement, nor of the other bases that make up the genetic code. That doesn't mean that they are easy for us to build into that vital configuration, as John Sutherland's work

demonstrated in 'The Origin of Life' (page 95), but those limitations are ours and not nature's. At some point in the distant past these chemicals acquired meaning and began to arrange themselves into a system from which information could be derived, stored and handed down. The system arises from the shape of the molecules – how they link with each other and form a chain. And we can now make our own chemicals, similar enough to these naturally occurring ones that they can be integrated into DNA. We're already using them in the fight against virus infections.

Typically, a virus will storm into a cell carrying its own genetic code, but none of the mechanics to translate it. One of the reasons viruses are not typically classed as living is because they can only reproduce by using someone else's kit. So viruses hijack a living cell, insert their own code and hope that it doesn't notice. This steganography by the virus uses the same language, so cells often don't spot the inveigled plan and read it unwittingly. As a result, the host cell produces more viruses, which break out to infect more cells, often destroying the host in the process. Antiviral treatments include false letters that are similar enough to be incorporated next to their natural counterparts, but conspicuous enough to disrupt any meaningful message. If you spot a massive interloping typo in the middle of a sent&ence, your brain is sophisticated enough to skip over it without misunderstanding. The cell is not so forgiving of the alien letters. They are designed so that the cell's translation mechanics see the new bases as unnatural, and won't get over them. This is how many antiviral and cancer chemotherapy drugs work. These man-made alien bases are used as standard treatments for infections from HIV/AIDS to herpes: when you finger a dab of cream onto your lips to soothe a cold sore you are probably applying acyclovir. The balm contains a molecule that is similar enough to the base G to go into the replicating virus code, but different enough to stop the sentence from continuing. The usefulness of this Scrabble-bag of alien letters comes from the very fact that they are alien. They don't act as a readable code. Instead

their insensibility to cellular machinery gives them power to rudely thwart the virus' sole desire – to copy itself. Reproduction is halted, making it a very useful tool for fighting these invaders.

Of course, the most interesting things we can do with invented DNA rely on our creating code that *can* be copied. If a code can't be copied, then it can't be passed on, and if it can't be translated, it can't have a function (beyond the terminating one described above). Many attempts have been made to make letters that can be snuck into DNA, and not only stay there, but get copied. Synthetic biologist Steve Benner has led this effort and, in 2011, added two alien bases to the alphabet of genetics.

The double helix of DNA has two spines that twist into a spiral, so it has two grooves on the outside – a helter-skelter with twin slides. One is wider than the other, so they are called the major and minor grooves. Electrons whizz around the atoms down in the minor groove, just as they do in all molecules. Here they form an electronic pattern that acts like a chemical identifier, and they play an essential role in DNA replication. The protein that copies genetic code, DNA polymerase, has evolved to recognize this particular electronic signature, and is drawn in to begin its copying work. In order to trick the cell into incorporating unnatural bases into DNA, Steve Benner's tactic was to design new ones, called Z and P, that mimic the electron signature when integrated into the double helix, and DNA polymerase gets into the groove. So now we have a DNA whose code is made up of an alphabet consisting of A, T, C, G, Z and P. DNA polymerase happily does its natural thing, reading the code and copying it. As always, the copying process is not perfect, and there is an error rate which is equivalent to the accumulation of the mutations in DNA that drive evolution. Benner has added brand new letters to the genetic code.

These new letters don't mean anything yet. This design and construction is a proof of principle – that extra letters can be added to DNA, and that they will not disrupt the cell's normal behaviour. The new techniques of biological engineering tend to rely on the existing mechanics of the cell to translate and enact the new

designs of synthetic biologists. Of course, alphabet letters don't have any inherent meaning in isolation either. It's only when strung together into consensus agreed words and phrases that they stop being mere noise and transform into prose that can express love and render the complex understandable. If we were to simply add new letters to the English alphabet, say Љ and Ц (from Cyrillic), we would have no way of pronouncing them, and they would not have any meaning until a consensus was agreed. But this is not the only way to invent new languages.

Alien Code

This focus on the letters of the spiralling ladder of DNA is only one aspect of synthetic biology's attempts to re-engineer the alphabet of life. DNA is a complex molecule, a polymer, meaning that it is assembled from repeating units. The code is hidden in the rungs of the ladder, but another major breakthrough in alternative, unnatural genetics has come not from swapping out or inventing new rungs, but the uprights of the ladder. These are made from a type of sugar called deoxyribose – the D in DNA (and in RNA it is simply ribose), which repeat and link up to form the spines of each strand. Since 2000, biologists have been creating alternative spines from a range of other sugars to produce several new genetic molecules: ANA (with arabinose), TNA (with threose) and four others such as FANA and CeNA. These alien species are collectively known as xeno-nucleic acids, or XNAs. The letters of code remain the same, but the struts of the ladder are strange. In April 2012, again at the Laboratory of Molecular Biology in Cambridge, a team led by Philip Holliger and Vitor Pinheiro built a system that for the first time enabled these foreign-language genes not only to replicate, but to evolve.

The genius of this experiment was not how the code was put together but the fact that it gets copied. DNA polymerase, the protein whose job is to copy a single strand of DNA to make it

into a double, does so by reading one strand and picking up and stringing together the pieces to make that strand's mirror, A for T, C for G and so on. DNA polymerase won't attach to anything other than DNA, understandably, as it has evolved over billions of years to do only that, and in nature there is nothing for it to act on but DNA. Proteins such as DNA polymerase are long and complicated, and their function is determined by their three-dimensional shape (which is determined by the order of their amino acids, which is determined by their genetic code). Because of this intricate 3D conformation, they resist intelligent design: it is immensely difficult to redesign something so complex. Evolution is blind but clever; it designs by random experimentation and by selection of successful mutants. Pinheiro presided over a process of natural selection by creating a pool of polymerase proteins that were all fractionally mutated in key areas. By selecting mutants that could pick up XNA pieces instead of DNA he forcefully evolved a polymerase that reads DNA, but lays down a mirror strand using XNA, as if the alien molecule was native. In this way, the brand new, human-invented genetic molecule can carry the same message encoded in natural DNA. They effectively tricked the machinery of cellular cryptography into reading its natural language but copying it into an unnatural one. They built a tool that will translate a natural language into a made-up one, English into Klingon.

But that is only half the story. They can also translate back again. Pinheiro mutated and selected a protein that would do the opposite, translate from XNA back to DNA. That is a trickier trick, as no natural protein will do the job. After eight rounds of forced random mutation and selection they had crafted something that did. That makes XNA a system that bears two hallmarks of life: information storage and heredity. Fidelity is important when copying, especially in genetics, but perfect fidelity is not evolution, it is stasis. Their process is faithful at around the 95 per cent mark, which means the stored information can evolve.

This, if one cares about such naming conventions, marks the

inception of a new branch of biology – 'synthetic genetics'. Admittedly this first study still relies on DNA as its template, but the team is already trying to cut it out of the loop. They have copied FANA to FANA and CeNA to CeNA, though it doesn't work quite as well as DNA yet.

The implications are weighty. Pinheiro and his colleagues have shown that genetic evolution is not limited to the natural code as we know it. There may well be uses for this gallimaufry of XNAs in the future, as they can behave like natural genetic parts, even though they are not, and might not be treated as such by our immune systems, which have several billion years' worth of schooling in recognizing natural, DNA- or RNA-based genetics. XNAs are more robust than DNA and RNA, both of which are prone to breakages and cuts by proteins called nucleases, whose role is to do just that. By being immune to slicing, XNAs have extraordinary therapeutic potential. Geneticist and origin-of-life scientist Jack Szostak has developed a class of molecule called an 'aptamer', which are strands of DNA or RNA that are designed to fold in such a way as to clasp to a very specific target. As a potential therapy, this action has the ability to shut down a gene, or inactivate a mutant protein. Currently there is only one aptamer on the market, as a treatment for a degenerative eye disease. But it is vulnerable to being chopped up by nucleases patrolling for rogue bits of DNA, and so the medicine has to be taken repeatedly. Theoretical at present, an equivalent XNA aptamer would be invisible to nuclease, a stealth weapon in the armoury of medicine.

Altered Output

Both of these projects reinvent the code of genetics. There is, of course, the output of that code to consider as well. Life is built by or of proteins, and with the advent of sophisticated molecular biology, proteins are now also subject to rewriting.

Proteins are built from the nose-to-tail joining of molecules

called amino acids. At one end an amino acid will have an arrangement of atoms called an amine group (which is made up of a
nitrogen atom and two hydrogens), and at the other a group that
forms carbolic acid (which is a carbon, two oxygens and a hydrogen). The genetic code in DNA spells out twenty amino acids,
which are collected together in the process of protein manufacture
and assembled by the cell.★

'Amino acid' is a generic name for a set of molecules that are all
quite similar. They vary only in what is attached to the middle section of the molecule, between the amine and the carbolic acid.
The simplest amino acid is glycine, with only a single hydrogen
atom as its unique side-chain. At the other end of the scale is tryp-
tophan, with a large double ring of carbons, hydrogens and a
nitrogen sprouting from the side. All manner of variations in
between make up life's lexicon. These side-chains determine the
behaviour of the protein in which the amino acids are assembled.
It's important to remember that there's no real limit to the number
of amino acids that could exist, in theory, beyond the twenty life
actually uses.

To get a sense of how this system has been breached by scientists
in the last few years, we first need to understand how the machinery of the cell works. Protein construction is remarkably like a
mechanical production line. The gene being translated is first transcribed from DNA into a free-floating copy in RNA, maybe a
thousand bases long, untethered to the rest of the DNA of the
host cell. This RNA message drifts to one of many ribosomes, the
centre where the protein will be constructed. In the cellular milieu
around the ribosome float the amino acids that will be strung
together according to the order set out in the messenger (or m)
RNA, each one specified by a three-base codon – the specific

★ In fact, our bodies make eleven of these amino acids, and we need to ingest a
further nine. There are two other amino acids that life forms infrequently use,
one of which is exclusive to bacteria and archaea. In general, we tend only to
refer to the canonical twenty.

triplet of letters that encodes an amino acid. Several key molecular players shift these molecules around in the manner of a production line, and the process works as follows. The gene to be translated (in the form of mRNA) is fed into the middle of the ribosome, like feeding paper ribbon into a tickertape machine. The RNA bases are read codon by codon, and the corresponding amino acids are added, one at a time. The protein is ejected like a tickertape being spat out. This output is the basic protein, to be folded and transported to its site of use. But the agents in this process are effectively manufactory handlers. The ribosome itself is of fundamental importance: its shape and construction, largely out of RNA itself, is the forge into which all the ingredients trundle along this biological conveyor belt. But also, there is a complex of three molecules that very specifically deliver the amino acids to the ribosome. This combination is basic to all life, but not easy to explain. One part is the amino acid itself, which is collected by the second part, a folded piece of RNA that specifically collects and transfers it to the ribosome. For this reason, these are generically called 'transfer (or t)RNA', and work by bearing the 'anti-codon', that is, bases complementary to the codon itself – a T to pick up an A, a C to pick up a G, and so on (though there are only twenty amino acids, there are dozens if not hundreds of tRNAs). The final handler is a protein with the unwieldy name of aminoacyl tRNA synthetase, which I shall refer to as synthetase henceforth. This protein clamps the amino acid to its corresponding tRNA, and together the component parts of the protein are delivered for assembly. The code input goes in at one end, the ingredients are delivered from another as a package, and the protein rumbles out, one amino acid at a time.

This process occurs in all life forms and its universality is one of the cornerstones of the evidence for a unique origin of life. The components are clearly ancient – no matter how distantly related two species are, they share the basic workings of this assembly plant. To change it has required an example of evolutionary brute

force, an intelligent design pioneered by Peter Schultz at Scripps in California, and by Jason Chin and his team at the LMB in Cambridge.

The existing code has plenty of redundancy built into it: there are sixty-four possible combinations of the four bases, and sixty-one of them are used to encode the twenty amino acids of life.* That leaves three remaining possible combinations, and in life forms all of them bear the message STOP, that is, the instruction to terminate protein production. These punctuations are essential to indicate the end of a gene. Indeed the significance of the stop codon is demonstrated in a handful of genetic diseases. Sandhoff disease, lethal in infancy, is caused by a gene bearing a premature stop codon, like a sentence cut incomprehensibly short by a rogue full stop.

One of them, the triplet UAG, is called the 'amber stop'.† Jason Chin has appropriated this sequence as a codon for building amino acids otherwise unrecognizable to living cells. There is no tRNA-synthetase pairing that will bind to an amber stop, and so Chin's idea is to evolve one. They start with this pairing from a species distantly related to the experimental cell. This ensures that the pair won't start interacting with the cell's normal workings: the mechanics of protein production are universal, but that doesn't mean that the kit is exactly the same. But in order to get the alien synthetase to singularly recognize something that none of the other tools of life can – the amber stop – they use the basic premise of evolution. If, for some reason, you wanted a fish that only ate something artificial such as jelly babies, you would start with a densely populated pool and use a jelly baby as bait. If there was a fish in that pool that, unusually, had a natural taste for jelly babies, you would catch it with that bait. You could then breed a new

---

* For example, six different combinations encode leucine: UUA, UUG, CUA, CUG, CUC and CUT; see 'The Origin of Life', Chapter 5 for the reasons for the evolution of this redundancy.

† Amber stop is not named in relation to the intermediate warning at traffic lights, but because its discoverer's name was Bernstein, the German for 'amber'.

pool from that fish and create a population that liked jelly babies. By repeating the process iteratively, you would end up with fish that only fed on jelly babies.

The fishing exercise is similar to the process of evolving a tRNA-synthetase pair that will recognize unnatural amino acids. By starting with a pool of versions of synthetase that have randomly mutated parts in the region that locks onto the amino acid, you can select ones that bind to your new unnatural amino acid, and repeat the process. Eventually, just like the jelly-baby-eating fish, you have created a tRNA-synthetase that will only read a stop and an unnatural amino acid.

That's one part of it. The tRNA is bred in that way to act as the anti-codon for the amber stop. Adding a UAG into the sequence of a gene so it can be picked up by the tRNA is fiddly, but essentially uses the tools of DNA manipulation not unlike those used for any gene tinkering – the creation of Freckles the spider-goat, for instance. You can synthesize a whole gene sequence, rewritten from scratch complete with an amber stop at a crucial point. Or you can use a copying technique, called mutagenic PCR, that will copy imperfectly, and introduce the new triplet in error. UAG normally only exists in a gene to indicate its end, so in order for this system to work, all the elements have to be changed: the code, the synthetase, the tRNA, all to recognize an unnatural amino acid.

This is delicate biological engineering to the most fundamental living process – the 'central dogma' as Francis Crick described it. But it is not mere experimentation for the sake of it. Jason Chin and others are using it to discover how proteins interact with each other. Living cells are a network of protein interactions: some attach themselves to DNA, some to the molecules of metabolism, and frequently they connect with other proteins to perform their vital functions. The unnatural amino acid incorporated into Chin's experiments carries a side-chain which does two things. First, when instructed to do so, it will form a locked bond with the protein it is interacting with, which means that we can fish the pair of them

out, and identify the unknown partner, the engineered protein acting as bait. And second, the unnatural amino acid is light-activated. To get it to lock onto the target, you merely shine ultraviolet light onto the cell, and the bond is sealed.

Chin is not the only person who can perform this act, but his team is the first to get it into multiple species, including animals. Until 2012, these unnatural actions were restricted to cells in culture, in dishes where the environment is controlled and neat. Chin has incorporated this process into fruit flies, one of the standard animals of experimental biology, and shown that no aspect of biology is off limits when it comes to re-engineering: alphabet, code, proteins – the whole system of life is now ripe for rewriting.

These rewriters show that, while our system, the only one we know of, is powerful enough to form the basis of all life on Earth, it doesn't have to be that way. Evolution by natural selection was discovered and described a long while before we understood its mechanics. The code that enacts that process of error and trial is unique, but is quite capable of being adjusted, rewritten or even invented afresh. It's difficult to conceive of a system of biological evolution (that doesn't rely on some form of supernatural creator) other than natural selection, but we only have one system of coded, reproducible information that has worked. XNA, the bases Z and P, and amino acids that are unnatural to the cell all show that there can be others. This is the first baby step towards the creation of new life forms that use a Darwinian genetic code that is invented not by nature, but fully by us.

Digital DNA

With these examples we can see how the encroachment of synthetic biology on nature has given us the power to radically alter the genetic code of DNA, to invent new versions of genetic material, and even to expand the lexicon of genetics so that it can include unnatural proteins.

The language of DNA is not restricted to the creation of life. Our genomes are data-storage devices, imperfect by natural design to encourage adaptation to the changing environment. But at the same time, DNA is remarkably stable. It has transmitted information in the same form for billions of years. And the information held in DNA remains intact long after the death of the cell or organism that bears it. More and more frequently these days stories emerge from the new scientific world of ancient DNA. The genome of the woolly mammoth has been published – extracted and decoded from hairs 60,000 years old, bought on eBay by the geneticists who sequenced it. Our likely evolutionary cousins the Neanderthals joined the genome club in 2010 when their complete DNA was read from 44,000-year-old bone specimens. But the current record stands with DNA extracted from frozen mud cores in Greenland, identified as being from the family of perennial plants saxifrages, and with an age range of between a whopping 450,000 and 800,000 years old. True, being buried beneath two kilometres of ice is about as close to a laboratory freezer as nature can manage, but the fact that we can find long-dead tissue and still decode the messages written within it shows what a stable data storage device DNA can be.

This is one of the reasons that scientists in the 1990s began thinking about DNA as a means of storing not just the biological information required for a cell to function, but digital data. Although not the first to take advantage of DNA's data-storage talents, Craig Venter's synthetic bacteria, aka Synthia, is the most famous living digital device to date. As mentioned in the Introduction, within the computer- and machine-synthesized genome of the goat pathogen *Mycoplasma mycoides*, Venter hid messages in the DNA that would be cryptic even to the machinery of the cell. Computer coders often hide secret treasures or messages – Easter eggs – in their programs to give credit, or simply as fun puzzles for users to discover.

There were four parts to these DNA Easter eggs. The first was a code table. The cryptic messages were all conceived in English,

and so a cipher had to be included, so that the four letters of genetic code could spell out the twenty-six letters of the English
alphabet plus punctuation. The twenty amino acids are sometimes
referred to by a single letter, A to W, so a conceivable way of con-
cealing an English language message in DNA would be to simply
use the existing genetic code in the triplets that spell out amino
acids, three bases for each Roman letter. But instead Venter constructed a new cipher for the English alphabet, and so this needed
to be deciphered. The code was cracked within a few days of
publication.

The second and third concealed messages comprised the names
of the few dozen creators of the cell, and an Internet address. The
final Easter egg was a set of three apposite quotations. The first
was 'to live, to err, to fall, to triumph, to recreate life out of life'
from the novel *Portrait of the Artist as a Young Man* by James Joyce .*
The second quotation came from J. Robert Oppenheimer, the so-
called father of the atomic bomb: 'See things not as they are, but
as they might be.' And the third was a fractional accidental mis-
quotation of that marvellous phrase from Richard Feynman,
reproduced in the Introduction: 'What I cannot build, I cannot
understand.'†

---

* As mentioned on p. 8, it was the subject of a copyright challenge by James
Joyce's estate.

† Which, in case you are wondering, looks like this in DNA : TTAACTAGCTAA
TTTCATTGCTGATCACTGTAGATATAGTGCATTCTATAAGTCGC
TCCCACAGGCTAGTGCTGCGCACGTTTTTCAGTGATATTATCC
TAGTGCTACATAACATCATAGTGCGTGATAAACCTGATACAATA
GGTGATATCATAGCAACTGAACTGACGTTGCATAGCTCAACTG
TGATCAGTGATATAGATTCTGATACTATAGCAACGTTGCGTGAT
ATTTTCACTACTGGCTTGACTGTAGTGCATATGATAGTACGTC
TAACTAGCATAACTAGTGATAGTTATATTTCTATAGCTGTACAT
ATTGTAATGCTGATAACTAGTGATATAATCCAACTAGATAGTCC
TGAACTGATCCCTATGCTAACTAGTGATAAACTAACTGATACAT
CGTTCCTGCTACGTGATAGCTTCACTGAGTTCCATACATCGTC
GTGCTTAAACATCAGTGATAACACTATAGAGTTCATAGATACTG

These blocks of DNA sequence were designed by hand and assembled in a computer. They had to be flanked by DNA tags that indicated that they were not to be translated by the cell, because the code they are written in is entirely new and would have been nonsense. That sequence is not written in the triplet codons of the genetic code, and cannot and will not make a protein. Instead, Venter had designed a storage device that has an encrypted code that was invented and new, and simply used the alphabet of DNA as code, without reference to its history.

In 2012, Harvard's George Church drove the commoditization of DNA into a new era with the first publication of an entire book encrypted digitally in DNA (called *Regenesis*, and appropriately enough it is about synthetic biology). At 53,000 words, eleven images and one script of software code, it would be three-quarters the size of the book you are holding, or around five megabits of information. The code is very simple binary: an A or C represents a 1, and a T or G represents a 0. Words were converted into a digital form and then the equivalent sequence of DNA synthesized in a computer in ninety-six base lengths, each with information tags about the location of the segments and so on – what coders call metadata. 55,000 of these fragments make up the book, roughly

---

```
CATTAACTAGTGATATGACTGCAAATAGCTTGACGTTTTGCAG
TCTAAAACAACGTGATAATTCTGTAGTGCTAGATACTATAGAT
TTCCTGCTAAGTGATAAGTCTACTGATTTACTAATGAATAGCT
TGGTTTTGGCATACACTGTGCGCTGCACTGGTGATAGCTTTT
CGTTGATGAATAATTTCCCTAGCACTGTGCGTGATATGCTAGA
TTCTGTAGATAGGCTAAATTCGTCTACGTTTGTAGGTGATAGT
TTAGTTGCTGTAACTAATATTATCCCTGTGCCGTTGCTAAGCT
GTGATATCATAGTGCTGCTAGATATGATAAGCAAACTAATAGA
GTCGAGGGGGAGTCTCATAGTGAATACTGATATTTTAGTGCTG
CCGTTGAATAAGTTCCCTGAACATTGTGATACTGATATTTTAG
TGCTGCCGTTGAATATCCTGCATTTAACTAGCTTGATAGTGCA
TTCGAGGAATACCCATACTACTGTTTTCATAGCTAATTATAGG
CTAACATTGCCAATAGTGCGGCGCGCCTTAACTAGCTAA.
```

one per word. In translating the whole book into DNA, there are only ten errors in five million bits of data. The whole thing is stored on what's called a DNA chip, which as a means of storing a book is not that different from a page and ink, just much smaller. These chips are small glass plates, about the size of a book of matches, coated with chemicals that DNA will stick to. The DNA data fragments are literally sprayed by a tiny nozzle from an inkjet printer. Unlike the information stored in Craig Venter's synthetic bacteria, Church's method never goes near a life form. The DNA is written in a computer, synthesized by a machine, printed with a printer and decoded by a machine and software using the very same standard techniques employed by scientists recovering DNA from millennia dead tissue.

Like many of the invented techniques in this chapter, this is a proof of principle. As a means of information storage, DNA's resilience is irrefutable. George Church and others have taken the simple fact that DNA can carry a code and have rewritten it to store non-biological data. The stability of DNA and the falling costs of the technology used gives this technique potential for archiving data at a breathtaking density: the authors claim 5.5 peta-bits per cubic millimetre (that is one quadrillion units of information). That makes this a more concentrated form of data storage than a Blu-Ray disc, a Flash drive or even the hard drive on your computer. What has been achieved so far is only useful for archiving, as the technology to write and access the memory stored in DNA takes many days, compared to elec-trical circuitry in a computer, which will save an equivalent amount of data in seconds. But it is not inconceivable that one day computers might store their memories not in silicon chips, but in DNA chips.

That creative engineering philosophy is key to synthetic biol-ogy: how can we redesign and use biological technology for purpose? All of the endeavours described above are fledgling tech-nologies, still very much in the experimental phase. But they all

demonstrate that the limits of nature are being surpassed by our own invention. We have always adapted nature for our benefit, and in the era of molecular biology we have done so by remixing it at a molecular level. Now, for the first time, we are engineering living systems that are created in ways that rewrite the very language that evolution has provided.

References, Notes and Further Reading

Introduction

Craig Venter's paper on building a cell from a fully manufactured genome: Daniel G. Gibson et al., 'Creation of a Bacterial Cell Controlled by a Chemically Synthesized Genome', *Science* 329 (2 July 2010), pp. 52–6 (doi:10.1126/science.1190719).

And the patent for the minimal genome of *Mycoplasma micoides*, or 'Synthia', as it was nicknamed: http://1.usa.gov/QsssVZ.

Venter sandwiched between David Cameron and Sarah Palin: *New Statesman*, 21 September 2010.

Chapter 1: Created, Not Begotten

I first physically encountered Freckles the spider-goat (and her creator Randy Lewis) during the filming of an episode of the BBC television science series *Horizon*. Parts of this chapter are taken from interviews from that filming, and from an article I wrote for the *Observer* newspaper based on that show. 'Synthetic Biology and the Rise of the "Spider-goats"'(Saturday 14 January 2012).

F. Teulé et al., 'A Protocol for the Production of Recombinant Spider Silk-like Proteins for Artificial Fiber Spinning', *Nature Protocols* 4:3 (2009), pp. 341–55 (doi:10.1038/nprot.2008.250).

Here is Hamilton Smith's 1970 paper that characterizes the restriction enzyme HinDII; it was this technology which enabled all molecular biology that followed: H. Smith and K. W. Wilcox, 'A Restriction Enzyme from Hemophilus Influenzae 1. Purification and General Properties', *Journal of Molecular Biology* 51 (1970), pp. 379–91 (doi:10.1016/0022-2836(70)90149-X).

Chapter 2: Logic in Life

The most comprehensive analysis of modern biotechnology is *Biology Is Technology: The Promise, Peril, and New Business of Engineering Life* by Robert H. Carlson (Harvard University Press, 2011), and is a brilliant primer for the science, economics and politics of this ever-changing field.

Ron Weiss's cancer assassin circuit: Zhen Xie et al., 'Multi-Input RNAi-Based Logic Circuit for Identification of Specific Cancer Cells', *Science* 333 (2 September 2011), pp. 1307–11 (doi:10.1126/science.1205527).

The cancer cells that are targeted and destroyed in this system are called HeLa cells. Due to their unlimited ability to divide – 'immortality' – they have been experimented on and distributed innumerable times, and are of immeasurable importance to many aspects of modern biology. They were derived in 1951 from the cervix of Henrietta Lacks, a poor black American who died of her cancer and lies in an unmarked grave. Her tale is told with grace and brilliance by Rebecca Skloot in the book *The Immortal Life of Henrietta Lacks* (Macmillan, 2010).

The two back-to-back papers that launched the component nature of genetic engineering at the birth of synthetic biology: Michael B. Elowitz and Stanislas Leibler, 'A Synthetic Oscillatory Network of Transcriptional Regulators', *Nature* 403 (20 January 2000), pp. 335–8 (doi:10.1038/35002125); Timothy S. Gardner, Charles R. Cantor and James J. Collins, 'The Construction of a Genetic Toggle Switch in *Escherichia coli*', *Nature* 403 (20 January 2000), pp. 339–42 (doi:10.1038/35002131).

Ten years later, the circuits had become significantly more sophisticated, for example: Tal Danino, Octavio Mondragón-Palomino, Lev Tsimring and Jeff Hasty, 'A Synchronized Quorum of Genetic Clocks', *Nature* 463 (21 January 2010), pp. 326–30 (doi:10.1038/nature08753).

At *Nature*, my colleague Charlotte Stoddart made a short film of the astonishing synthetic synchronization of these bacteria and published it on YouTube. In it, Charlotte's voiceover describes the fluorescing action as in a 'Mexican wave', as it closely resembles the metachronal rhythm demonstrated in packed stadiums by sports fans standing, yelling and sitting down in succession. YouTube viewers are not known for providing the most sophisticated commentary below the videos themselves, but we were surprised to learn that the term 'Mexican wave' was primarily a British phrase, as a result of its occurrence at football (soccer)

matches in the Mexico World Cup, 1986. It seems that many other people in the world took this as a vaguely racist comment.

On the dangers of a trip to Mars: W. Friedberg et al., 'Health Aspects of Radiation Exposure on a Simulated Mission to Mars', *Radioactivity in the Environment* 7 (2005), pp. 894–901.

A good analysis of the claims and hype of synthetic biology: Roberta Kwok, 'Five Hard Truths for Synthetic Biology', *Nature* 463 (20 January 2010), pp. 288–90 (doi:10.1038/463288a).

Chapter 3: Remix and Revolution

My thoughts on copying, creativity and music in relation to evolution and synthetic biology are influenced, appropriately enough, by a series of online videos by New York filmmaker Kirby Ferguson called 'Everything Is a Remix', available in four parts at http://www.every thingisaremix.info/.

Completing the loop, Ferguson used evolution as an introductory metaphor for the genesis of ideas in Part 4 of his outstanding piece of work, which he formulated with a very small amount of help from me. I now gleefully extend this idea into the realm of synthetic biology by borrowing and transforming his ideas. I'm pretty sure he won't mind.

The changing nature of creativity, specifically in music, as a result of copyright law is excellently documented by Joanna Demers in *Steal This Music* (University of Georgia Press, 2006).

One of the most striking examples of sampling in music is the innumerable reuses of a short stretch of drum solo from the B-side of a 1969 7-inch single by The Winstons, a largely forgotten funk soul band. The four-bar section is known as the 'Amen Break', as the song itself was called 'Amen Brother', and features a dropped and syncopated drum hit known as a 'breakbeat'. In the mid 1980s it began appearing as samples in tracks on the developing New York hip-hop scene. By the end of the 1980s, it had been sampled widely, by many dance and hip-hop bands, including in 'Straight Outta Compton' by the influential group NWA. A decade later, the seven-second drum loop had appeared on hundreds of songs ranging from a David Bowie single to the credits to the cartoon *Futurama*. And the entire genres of jungle music and drum and bass are based around the Amen Break. A database of tracks that use it can be found at http://amenbreakdb.com/.

The first album to be constructed entirely from samples is *Endtroducing* (1996) by DJ Shadow, using at least ninety-nine samples over the eighteen tracks, from films including *Silent Running* (1972) and *Prince of Darkness* (1987), and bands including Queen, Metallica, Pink Floyd, Kraftwerk and Nirvana. Craig Venter's bacteria nicknamed Synthia is probably the closest to a fully sampled living cell, though *Endtroducing* is more culturally relevant and much more exciting.

'Hello world': Anselm Levskaya et al., 'Synthetic Biology: Engineering *Escherichia coli* to See Light', *Nature* 438 (24 November 2005), pp. 441–2 (doi:10.1038/nature04405).

Three months of measuring the wings and eyes of several thousand stalk-eyed flies (*Cyrtodiopsis dalmanni*), after breeding and raising them in liquefied rotting sweetcorn – happy days: P. David, A. Hingle, D. Greig, A. Rutherford, A. Pomiankowski and K. Fowler, 'Male Sexual Ornament Size but Not Asymmetry Reflects Condition in Stalk-eyed Flies', *Proceedings of the Royal Society B* 265 (1998), p. 2211 (doi:10.1098/rspb.1998.0561).

The beginnings of the biocementation process that the Brown–Stanford iGem team would use to look at RegoBrick formation for terraforming: S. S. Bang and V. Ramakrishnan, 'Calcite Precipitation Induced by Polyurethane-immobilized *Bacillus pasteurii*', *Enzyme and Microbial Technology* 28 (2001), pp. 404–9.

Two useful reviews of patents in genetics and synthetic biology: Arti Rai and James Boyle, 'Synthetic Biology: Caught Between Property Rights, the Public Domain, and the Commons', *PLoS Biology* 5 (2007), p. e58 (doi:10.1371/journal.pbio.0050058); Berthold Rutz, 'Synthetic Biology and Patents: A European Perspective', *EMBO Reports* 10 (2009), pp. S14–S17 (doi:10.1038/embor.2009.131).

The European patent for the *Oncomouse*: http://register.epoline.org/espacenet/application?number=EP85304490.

President Obama's *National Bioeconomy Blueprint*: http://www.white house.gov/sites/default/files/microsites/ostp/national_bioeconomy_blueprint_april_2012.pdf.

Chapter 4: The Case for Progress

Martin Robbins, '"HULK SMASH GM" – Mixing Angry Greens with Bad Science', Wednesday 30 May 2012, guardian.co.uk.

Sophie Vandermoten, 'Aphid Alarm Pheromone: An Overview of Current Knowledge on Biosynthesis and Functions', *Insect Biochemistry and Molecular Biology* 42 (2012), pp. 155–63.

G. Kunert, C. Reinhold and J. Gershenzon, 'Constitutive Emission of the Aphid Alarm Pheromone, (E)-beta-farnesene, from Plants Does Not Serve as a Direct Defense Against Aphids', *BMC Ecology*, 10 (2010), p. 23 (doi:10.1186/1472-6785-10-23).

L. G. Firbank et al., 'Farm-scale Evaluation of Genetically Modified Crops', *Nature* 399 (1999), pp. 727–8.

J. N. Perry et al., 'Ban on Triazine Herbicides Likely to Reduce but Not Negate Relative Benefits of GMHT Maize Cropping', *Nature* 428 (18 March 2004), pp. 313 16 (doi:10.1038/nature02374).

M. S. Heard et al., 'Weeds in Fields with Contrasting Conventional and Genetically Modified Herbicide-tolerant Crops. 1. Effects on Abundance and Diversity', *Philosophical Transactions of the Royal Society B* 358 (2003), pp. 1819–32.

J. N. Perry et al. 'Design, Analysis and Power of the Farm-scale Evaluations of Genetically-modified Herbicide-tolerant Crops', *Journal of Applied Ecology* 40 (2003), pp. 17–31.

Chelsea Snella et al., 'Assessment of the Health Impact of GM Plant Diets in Long-term and Multigenerational Animal Feeding Trials: A Literature Review', *Food and Chemical Toxicology* 50 (2012), pp. 1134–48.

Take the Flour Back's website: http://taketheflourback.org/.

The US Presidential Commission's recommendations on synthetic biology, 2010: 'New Directions: The Ethics of Synthetic Biology and Emerging Technologies', http://bioethics.gov/cms/synthetic-biology-report.

And the response from Friends of the Earth and ETC: http://www.foe.org/news/blog/2010-12-groups-criticize-presidential-commissions-recommenda.

The Obama Administration's *National Bioeconomy Blueprint*: http://www.whitehouse.gov/blog/2012/04/26/national-bioeconomy-blueprint-released.

And the response from Friends of the Earth and ETC: 'Principles for the Oversight of Synthetic Biology', http://www.foe.org/projects/food-and-technology/blog/2012-03-global-coalition-calls-oversight-synthetic-biology.

This classic paper was the first real instance of genetic modification, by Paul Berg and his team: D. A. Jackson, R. H. Symons and P. Berg, 'Biochemical Method for Inserting New Genetic Information into DNA of Simian Virus 40: Circular SV40 DNA Containing Lambda

Phage Genes and the Galactose Operon of *Escherichia coli*', *Proceedings of the National Academy of Science USA* 69 (1972), pp. 2904–9.

The post-Asilomar schism reported: 'Environmental Groups Lose Friends in Effort to Control DNA Research', *Science* 202 (1978), p. 22.

ETC's analysis of the bio-based economy: 'The New Biomassters: Synthetic Biology and the Next Assault on Biodiversity and Livelihoods', http://www.etcgroup.org/content/new-biomassters.

The *Guardian*'s stunt to assemble parts of the smallpox genome by mail order: James Randerson, 'Revealed: the Lax Laws That Could Allow Assembly of Deadly Virus DNA', *Guardian*, Wednesday 14 June 2006.

The 2002 construction of the virus that causes polio by taking its genome sequence from publicly available databases: Jeronimo Cello, Aniko V. Paul and Eckard Wimmer, 'Chemical Synthesis of Poliovirus cDNA: Generation of Infectious Virus in the Absence of Natural Template', *Science* 297 (9 August 2002), pp. 1016–18 (doi:10.1126/science.1072266).

And of the 1918 flu: Terrence M. Tumpey et al., 'Characterization of the Reconstructed 1918 Spanish Influenza Pandemic Virus', *Science* 310 (2005), pp. 77–80 (doi:10.1126/science.1119392).

The two experimental flu papers, one by Kawaoka in *Nature* and the other by Fouchier in *Science*, both finally published in full in June 2012 after much deliberation and debate, and an earlier editorial from *Nature*: Masaki Imai et al., 'Experimental Adaptation of an Influenza H5 HA Confers Respiratory Droplet Transmission to a Reassortant H5 HA/H1N1 Virus in Ferrets', *Nature* 486 (21 June 2012), pp. 420–28 (doi:10.1038/nature10831); Sander Herfst et al., 'Airborne Transmission of Influenza A/H5N1 Virus Between Ferrets', *Science* 336 (22 June 2012), pp. 1534–41 (doi:10.1126/science.1213362); 'Publishing Risky Research', *Nature* 485 (3 May 2012), p. 5 (doi:10.1038/485005a).

Claire M. Fraser and Malcolm R. Dando, 'Genomics and Future Biological Weapons: The Need for Preventive Action by the Biomedical Community', *Nature Genetics* 29 (2001), pp. 253–6 (doi:10.1038/ng763).

Gilles-Eric Séralini's controversial study on the negative effects of GM crops in rats' diets: Gilles-Eric Séralini et al., 'Long term Toxicity of a Roundup Herbicide and a Roundup-tolerant Genetically Modified Maize', *Food and Chemical Toxicology* 50 (2012), pp. 4221–31.

And two months later, a damning review by the European Food Safety Authority: European Food Safety Authority, 'Final Review of the Séralini et al. (2012a) Publication on a 2-year Rodent Feeding Study

with Glyphosate Formulations and GM Maize NK603 as Published Online on 19 September 2012 in Food and Chemical Toxicology', *EFSA Journal* 10 (2012), p. 2986.

On the creation of artemisinin: Dae-Kyun Ro et al., 'Production of the Antimalarial Drug Precursor Artemisinic Acid in Engineered Yeast', *Nature* 440 (13 April 2006), pp. 940–43 (doi:10.1038/nature04640); P. J. Westfall et al., 'Production of Amorphadiene in Yeast, and Its Conversion to Dihydroartemisinic Acid, Precursor to the Antimalarial Agent Artemisinin', *Proceedings of the National Academy of Science USA* 109 E111-8 (12 January 2012); Declan Butler, 'Malaria Drug-makers Ignore WHO Ban', *Nature* 460 (14 July 2009), pp. 310–11 (doi:10.1038/460310b); Melissa Lee Phillips, 'Genome Analysis Homes in on Malaria-drug Resistance', *Nature News* (5 April 2012) (doi:10.1038/nature.2012.10398).

The UK's Chief scientific adviser on GM foods: 'GM Food Needed to Avert Global Crisis, Says Government Adviser', *Telegraph* (24 January 2011).

'An affront to God': 'Meanings of "Life"', *Nature* 447 (28 June 2007), pp. 1031–2 (doi:10.1038/4471031b).

Paul Berg, David Baltimore, Sydney Brenner, Richard O. Roblin III and Maxine F. Singer, 'Summary Statement of the Asilomar Conference on Recombinant DNA Molecules', *Proceedings of the National Academy of Science USA* 72 (1975), pp. 1981–4.

A major public assessment of how various publics view synthetic biology in the UK: 'BBSRC Synthetic Biology Dialogue', http://www.bbsrc.ac.uk/web/FILES/Reviews/synbio_summary-report.pdf.

Afterword

Babel, Esperanto, Klingon, Babm, Blissymbolics, Loglan and Lojba – Arika Okrent's wonderful book on the stories of how we have started nearly 900 new languages: Arika Okrent, *In the Land of Invented Languages: Esperanto Rock Stars, Klingon Poets, Loglan Lovers, and the Mad Dreamers Who Tried to Build a Perfect Language* (Spiegel & Grau, 2009).

Steve Benner's addition of Z and P to the natural bases A, T, C and G: Z. Yang and F. Chen et al., 'Amplification, Mutation, and Sequencing of a Six-Letter Synthetic Genetic System', *Journal of the American Chemical Society*, 2011 (doi:10.1021/ja204910n).

XNA and the birth of synthetic genetics: Vitor B. Pinheiro et al., 'Synthetic Genetic Polymers Capable of Heredity and Evolution', *Science* 336 (20 April 2012), pp. 341–4 (doi:10.1126/science.1217622).

Unnatural amino acids: Lloyd Davis and Jason W. Chin, 'Designer Proteins: Applications of Genetic Code Expansion in Cell Biology', *Nature Reviews Molecular Cell Biology* 13 (February 2012), pp. 168–82 (doi:10.1038/nrm3286); Jason W. Chin et al., 'Addition of a Photocrosslinking Amino Acid to the Genetic Code of *Escherichia coli*', *PNAS* 99 (2002), pp. 11020–24 (doi:10.1073/pnas.172226299).

On ancient DNA: one of the most surprising interviews I have done, for the Nature Podcast, was that with Stephan C. Schuster, the lead researcher on the sequencing of the mammoth genome, 20 November 2008:

> Schuster: Once we realized that there might be the possibility that the hair shaft contained actual DNA, we went out and tried to find sources of DNA that we could reliably step into. And the way I do this, I started searching for them in eBay and when I immediately found that there are zillions of hairs available, we contacted the seller and then, together with authorities of the university, we made sure that there were proper import permits, and we also had palaeontologists and museum curators in Russia to check on the sources of that because we were very worried that some of those hairs and fossils might have illegally been sold outside Russia. And after we verified that all of this was taken care of and there is a clean record, then we started buying hair in a larger supply from that source.
>
> AR: Some of these details aren't mentioned in the methods section of the paper. Can I just get you to say that again: you bought these hairs off eBay? What was the successful bid?
>
> Schuster: For a handful of hair, I think it cost me something like 132 bucks.

Webb Miller et al., 'Sequencing the Nuclear Genome of the Extinct Woolly Mammoth', *Nature* 456 (20 November 2008), pp. 387–90 (doi:10.1038/nature07446).

An assessment of sequencing DNA from long-dead specimens: S. Pääbo et al., 'Genetic Analyses from Ancient DNA', *Annual Review of Genetics* 38 (2004), pp. 645–79 (doi:10.1146/annurev.genet.37.110801.143214).

On very old human DNA: Richard E. Green et al., 'A Draft Sequence of the Neanderthal Genome', *Science* 328 (7 May 2010), pp. 710–22 (doi:10.1126/science.1188021).

Exceedingly old plant DNA: Eske Willerslev et al., 'Ancient Biomolecules from Deep Ice Cores Reveal a Forested Southern Greenland', *Science* 317 (6 July 2007), pp. 111–14 (doi: 10.1126/science.1141758).

One of the first suggestions in the academic literature that DNA could be used as a digital storage device: Eric B. Baum, 'Building an Associative Memory Vastly Larger Than the Brain', *Science* 268 (28 April 1995), pp. 583–5 (doi:10.1126/science.7725109).

The most advanced realization of DNA as a means of storing data: George M. Church, Yuan Gao and Sriram Kosuri, 'Next-Generation Digital Information Storage in DNA', *Science* 337 (28 September 2012) (doi:10.1126/science.1226355).

Index

Acknowledgements

Just like science, writing this book has been a collaborative process. I wrote the words, but *Creation* is a concatenation and remix assembled from the ideas and work of others.

My thinking on the subject of the origin of life is heavily influenced by my friend Nick Lane at University College London. I embarked on *Creation* before reading his outstanding book *Life Ascending*. But, having done so now, and for our many lunches, I am profoundly indebted to him. In the first half, I've liberally borrowed from his thoughts, which he expresses with virtuoso clarity. I urge you to read it.

Many of the interviews and much of the research for this book were drawn from my involvement in BBC radio and television productions. In 2009 I presented a documentary series on BBC4 called *The Cell*, which over three hours covered the story of biology from van Leuwenhoek to synthetic biology. This was made by a team of brilliant filmmakers including Jacqui Smith, Andrew Thompson, Alison Rooper, Nick Jordan and Jill Fullerton Smith. Particularly, David Briggs did significant research on the early and nineteenth-century microscopists, and deserves acres of credit for academically and meticulously collating multiple original sources. In 2010, we followed this up with a BBC Radio 4 *Frontiers* programme on the current status of origin of life biology. Roland Pease produced, and immense gratitude is due to him as well as Sasha Feachem and Deborah Cohen at the BBC Radio Science Unit. During the writing of *Creation*, I filmed an episode of the long-running BBC flagship science programme *Horizon* on synthetic biology. Matthew Dyas, Kelly Neaves and Aidan Laverty skilfully guided that into the world. Some of the interviews

quoted in *Creation* were conducted under the auspices of those programmes.

Here are some others to whom I owe a debt of gratitude: Jane Sowden, my friend and Ph.D. supervisor, from whom I learned how to be a good scientist; Steve Jones, for teaching me, inspiring me and appearing as a contributor in almost all of my TV and radio programmes; Armand Leroi, for helping me with Aristotle by showing me chapters from his book *Aristotle's Lagoon*; Matt Ridley, for providing scientific and biographical information on Francis Crick; Rob Carlson, author of the excellent book *Biology Is Technology*, for his help on many aspects of synthetic biology; Alex Taylor and Vitor Pinheiro, for much-needed guidance on XNA and expanded genetic code; John Sutherland, for RNA synthesis; Zita Martins, for her expertise on astrobiology and Martian rocks; Emma Perry, for helping me with the school curriculum; Sinjoro Simon Varwell, dankon; those on Twitter who have at times responded to my pleas to find the right word, fungus or reptile, specifically @mr_muse, @sarcasmaniac, @janieface1, @writerJames; Kirby Ferguson, for copyright and remixing; Richard Kelwick, for the UK iGEM meet-up; Philip Campbell, Kerri Smith, Charlotte Stoddart, Geoff Marsh and Thea Cunningham, all my friends at *Nature*.

I was nomadic in writing, though rarely ventured out of Hackney. The following kind people allowed me to use their various spaces: Jack and Lynsey Mathew, and Barney for keeping me company (Lynsey also for the word 'jumentous'); John Sanders and Beth Gibbon; Ana-Paula Lloyd and Jonny Hassid; and everyone at my second home, Mouse and Delotz – quite clearly the best café in London.

Several people read and commented on the manuscript during its lengthy gestation. However, any errors or mutations that have survived these rounds of selection are mine alone. I wish to thank Ed Yong, Jim Al-Khalili, Matt Ridley, Dara O Briain, Douglas Adams and his family, Matthew Cobb, Brian Cox, Suzi Gage, David Adam, Nathaniel Rutherford and David Watson for their

advice, conversation and edits, and particularly Kevin Fong, with whom I have shared the chronic condition of writing. I tested many of the ideas in this book on him.

I am tremendously grateful to my agents, Sophie Laurimore and Will Francis; without Will's continual support *Creation* could never have happened; my wife Georgia and our children have supported its genesis unflinchingly, and my love for them is as deep as geological time. Most of all, I humbly thank my editor at Viking, Will Hammond, who has wrestled with me and these words countless times to batter them into shape. I doubt many writers receive as much attention in sculpting their prose.

Scientists can be a quarrelsome lot. They compete for ideas, and for funding, and, just like in any human endeavour, are subject to dogma and whimsy and hard-felt beliefs. Good theories emerge from that maelstrom of open – but sometimes fractious – debate, and get tested and refined by experiment. Science does not settle on truth, but tends towards it. These particular areas of science are pioneering and important and, as such, they are subject to quarrel perhaps more than most. The ideas presented here will be refined or changed over time, or rejected. And frankly, like all good scientists should, I look forward to being corrected.